Advances in Industrial Control

Springer-Verlag London Ltd.

Other titles published in this Series:

Feedback Control Theory for Dynamic Traffic Assignment
Pushkin Kachroo and Kaan Özbay

Robust Aeroservoelastic Stability Analysis
Rick Lind and Marty Brenner

Performance Assessment of Control Loops: Theory and Applications
Biao Huang and Sirish L. Shah

Advances in PID Control
Tan Kok Kiong, Wang Quing-Guo and Hang Chang Chieh with Tore J. Hägglund

Advanced Control with Recurrent High-order Neural Networks: Theory and Industrial Applications
George A. Rovithakis and Manolis A. Christodoulou

Structure and Synthesis of PID Controllers
Aniruddha Datta, Ming-Tzu Ho and Shankar P. Bhattacharyya

Data-driven Techniques for Fault Detection and Diagnosis in Chemical Processes
Evan L. Russell, Leo H. Chiang and Richard D. Braatz

Bounded Dynamic Stochastic Systems: Modelling and Control
Hong Wang

Non-linear Model-based Process Control
Rashid M. Ansari and Moses O. Tadé

Identification and Control of Sheet and Film Processes
Andrew P. Featherstone, Jeremy G. VanAntwerp and Richard D. Braatz

Precision Motion Control
Tan Kok Kiong, Lee Tong Heng, Dou Huifang and Huang Sunan

Nonlinear Identification and Control: A Neural Network Approach
Guoping Liu

Digital Controller Implementation and Fragility: A Modern Perspective
Robert S.H. Istepanian and James F. Whidborne

Optimisation of Industrial Processes at Supervisory Level
Doris Sáez, Aldo Cipriano and Andrzej W. Ordys

Applied Predictive Control
Huang Sunan, Tan Kok Kiong and Lee Tong Heng

Hard Disk Drive Servo Systems
Ben M. Chen, Tong H. Lee and Venkatakrishnan Venkataramanan

Nikolaos Xiros

Robust Control of Diesel Ship Propulsion

With 55 Figures

Springer

Nikolaos Xiros, Dr-Eng
Department of Naval Architecture and Marine Engineering, Laboratory of Marine Engineering, National Technical University of Athens, PO Box 64033, Zografos, 15710, Athens, Greece

British Library Cataloguing in Publication Data
Xiros, Nikolaos
Robust control of diesel ship propulsion. - (Advances in industrial control)
1.Marine diesel motors - Automatic control 2.Ship propulsion - Automatic control 3.Robust control
I.Title
623.8'7'236
ISBN 978-1-4471-1102-3

Library of Congress Cataloging-in-Publication Data
A catalog record for this book is available from the Library of Congress.

ISBN 978-1-4471-1102-3 ISBN 978-1-4471-0191-8 (eBook)
DOI 10.1007/978-1-4471-0191-8

http://www.springer.co.uk

Originally published by Springer-Verlag London Berlin Heidelberg in 2002
Softcover reprint of the hardcover 1st edition 2002

Typesetting: Electronic text files prepared by authors
69/3830-543210 Printed on acid-free paper SPIN 10845389

Advances in Industrial Control

Professor Emeritus O. P. Malik
Department of Electrical and Computer Engineering
University of Calgary
2500, University Drive, NW
Calgary
Alberta
T2N 1N4
Canada

Doctor K.-F. Man
Electronic Engineering Department
City University of Hong Kong
Tat Chee Avenue
Kowloon
Hong Kong

Professor G. Olsson
Department of Industrial Electrical Engineering and Automation
Lund Institute of Technology
Box 118
S-221 00 Lund
Sweden

Professor A. Ray
Pennsylvania State University
Department of Mechanical Engineering
0329 Reber Building
University Park
PA 16802
USA

Professor D. E. Seborg
Chemical Engineering
3335 Engineering II
University of California Santa Barbara
Santa Barbara
CA 93106
USA

Doctor I. Yamamoto
Technical Headquarters
Nagasaki Research & Development Center
Mitsubishi Heavy Industries Ltd
5-717-1, Fukahori-Machi
Nagasaki 851-0392
Japan

SERIES EDITORS' FOREWORD

The series *Advances in Industrial Control* aims to report and encourage technology transfer in control engineering. The rapid development of control technology has an impact on all areas of the control discipline. New theory, new controllers, actuators, sensors, new industrial processes, computer methods, new applications, new philosophies..., new challenges. Much of this development work resides in industrial reports, feasibility study papers and the reports of advanced collaborative projects. The series offers an opportunity for researchers to present an extended exposition of such new work in all aspects of industrial control for wider and rapid dissemination.

As fuel becomes more expensive, as engine technology changes and as marine safety requirements become more stringent there is a continuing need to re-investigate and re-assess the controller strategies used for marine vessels. Nikolaos Xiros has produced such a contribution in this *Advances in Industrial Control* monograph on the control of diesel ship propulsion. The monograph is carefully crafted and gives the full engineering and system background before embarking on the modelling stages of the work. The physical system modelling is then used to investigate both transfer function and state space models for the engine dynamics. This assessment yields a full appreciation of the need for a more detailed transfer function model in some operating regimes. However, when models are simplified, the requirement for robust control design emerges. In Chapter 4 such a robust PID control solution is indeed pursued along with the necessary steps to avoid implementing a D-term in the controller.

The last two chapters of the monograph examine state-space models and robust state-feedback control solutions. In this framework a more sophisticated control architecture is proposed and a more comprehensive control solution followed incorporating supervisory set point control.

Marine control problems continue to be of considerable industrial interest as evidenced by the strong support for IFAC's Control and Applications of Marine Systems (CAMS) events. Dr. Xiros has shown that a full understanding of marine engine physical systems is needed to construct suitable models and design appropriate controllers. The methodology in the monograph should be of interest to the wider control engineering and academic community whilst the detailed results will be of particular interest to marine control engineers and practitioners.

M.J. Grimble and M.A. Johnson
Industrial Control Centre
Glasgow, Scotland, U.K.

PREFACE

One of the most typical application paradigms, used widely in introductory control engineering textbooks, is the fly-ball (fly-weight) speed governor employed by James Watt for speed (rpm) regulation of the reciprocating steam engine he invented. The same type of engine, equipped with the same primitive control element, was used for ship propulsion in the early "steamers".

The same fly-ball system used by Watt in steam engines, was employed later in the 19th and 20th centuries for speed regulation of internal combustion engines and turbines. The functions incorporated in this device contain all the elements of a modern feedback control loop, integrated, though, in the same physical unit. There is a sensing element (sensor) and a negative-gain, error-amplifying mechanism that generates a driving signal for the hydraulic or mechanical power actuator of the unit. Although simple in its concept, this speed-regulating device remained in service until the end of 70s and 80s with some minor modifications, including the incorporation of electric circuitry for the generation of the actuator driving signals. However, progress in analogue and digital electronics made possible the development of electronic engine control units, which have been proven to be more reliable in service and flexible to cope with variable requirements and contexts of operation.

Electronic marine engine control has allowed for the direct implementation of the PID control law with gain scheduling. As the marine control engineers have got rid of the hardware and reliability limitations inherent in mechanical/hydraulic devices, the focus has moved to the control and regulation of the plant itself. The need of gain scheduling has been imperative, in the first place, as the combustion process in the engine cylinders is highly non-linear. Furthermore, as marine engines are turbocharged, an additional and variable time delay is introduced when the plant is accelerating or decelerating rapidly. Last, but not least, propeller loading introduces non-linearity, as well, and a significant amount of uncertainty and disturbance. It should be noted, however, that the marine propulsion system with fouled hull propeller law loading is an intrinsically stable system, from the control point of view. This is due to the dependence of the propeller load torque on shaft speed, which is monotonically increasing. Therefore, if for some reason the system equilibrium is disturbed, *e.g.* engine/propeller rpm is increased, a counter-effect, *e.g.* an increased value of propeller load, will decelerate the shaft.

However, although stability of the open-loop system is guaranteed, significant margins are introduced to a merchant ship's main engine, which eventually increase costs significantly. On the other hand, as explained in the text, engine over-sizing can be avoided if appropriate engine control is employed. In that respect, the subject of this text is to investigate PID and linear-state-feedback controller synthesis methods for achieving adequate disturbance rejection of propeller load fluctuation and robustness against parametric uncertainty and neglected dynamics.

As a state-space model of the system is required for the development of any state-space control design methodology, a way to derive such a model from the physical, thermodynamic engine description is given. This method is based on the non-linear mapping abilities of neural nets. Note that the value of the method is not limited to marine powerplant modelling, but can be employed in the case of any process or system where non-linear dynamics are present. The same holds for the controller synthesis methodologies proposed; although inspired by the robust control generic synthesis framework, they aim to simplify the mathematical intricacies of the formal method, provide an easier to manipulate form and, at the end of the day, make them more attractive to applications in the marine field or elsewhere. The methodology concerns SISO systems with PID control and 2×1 multivariable systems with full-state-feedback control and description available in state-space; additionally, it allows one to deal with robustness in a more intuitive way, as it is essentially a pole placement technique.

In conclusion, the text, although originally aimed at the field of marine powerplant control and regulation, I would hope to be of value to the control community as a whole, by providing additional insight into robust control design of processes and systems.

CONTENTS

LIST OF TABLES

CHAPTER 1

INTRODUCTION

1.1 The Marine Diesel Propulsion System

1.1.1 Historical Note

Propulsion of the vast majority of modern merchant ships (*e.g.* containerships and VLCCs) utilises the marine Diesel engine as propeller prime mover. Typical marine propulsion plants include a single, long-stroke, slow-speed, turbocharged, two-stroke Diesel engine directly coupled to the vessel's single large-diameter, fixed-pitch propeller. This configuration can reach quite large power outputs (up to 30–40 MW from a single unit) and yet is characterised by operational reliability due to its conceptual simplicity.

Since mechanisation of propulsion was first introduced in shipping in the mid-19th century various eras can be clearly distinguished. Early motor ships were propelled by side wheels or screw propellers and powered by reciprocating steam engines appropriately arranged in the vessel's hull. Later, transition to steam turbine powerplants was slowly effectuated and was completed by the end of World War II.

However, today the Diesel engine dominates over marine propulsion [1]. There are three major reasons for this fact [2,3]: (a) the superior (thermal) efficiency of Diesel engines over the other propulsion prime movers, (b) following the use of alkaline cylinder lubrication oils, large Diesel engines can burn heavy fuel oil (HFO) and (c) slow-speed Diesel engines can be directly connected to the propeller without the need of gearbox and/or clutch and are reversible. On the other hand, Diesel engines require a larger engine room compared to gas turbines, their major rival nowadays. Indeed, Diesel engines have lower specific power per unit volume and weight. This can be a problem when extremely large power outputs are required, *e.g.* for aircraft carriers or some projected large high-speed vessels.

1.1.2 Marine Engine Configuration and Operation

The propulsion demands of large merchant vessels can be covered using a single slow-speed, direct-drive Diesel engine. This type of engine can burn very low quality fuel, such as HFO, more easily than medium-speed Diesel engines because the physical space and time available to combustion are significantly larger. Slow-speed engines are usually built with a smaller number of cylinders and, in consequence, a smaller number of moving parts, increasing thus the reliability of the propulsion system. A section of a large marine, two-stroke, turbocharged engine used for ship propulsion is shown in Figure 1.1.

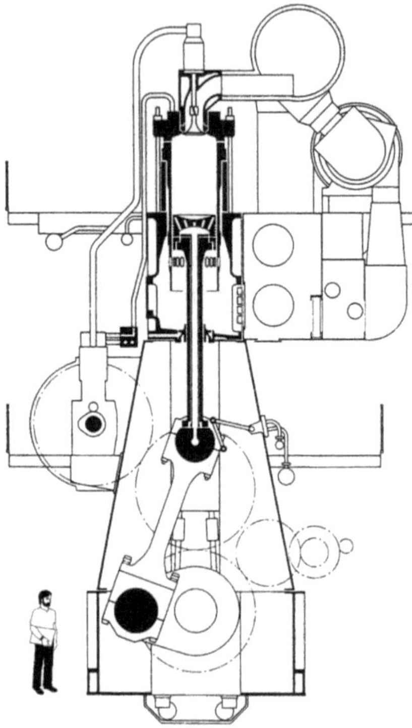

Figure 1.1 Section of large marine Diesel engine

The main parts of the engine are:

- Bedplate and crankcase: The bedplate supports the engine and ensures alignment of the shaft. The crankcase provides housing for the crankshaft. In large engines the crankcase and bedplate come in one piece.
- Crankshaft and flywheel: The crankshaft is one of the heaviest and costliest components of large marine engines. The flywheel ensures attenuation of the vibrations introduced by the discrete firings in each cylinder.
- Engine body: This part of the engine provides mechanical support for the engine cylinders and ensures the mechanical robustness and flexibility of the engine structure. Account is also taken that human access to the inner parts of the engine remains possible.
- Cylinder blocks and liners: In large marine diesels each cylinder is contained in it's own separate cylinder block.
- Pistons and connecting rods: Pistons consist of the piston crown, the piston rings and the piston rod. Their role is to deliver mechanical power

to the crankshaft to which they are connected through the connecting rods.

- Cylinder heads and exhaust valves: The cylinder heads secure the top of the combustion chamber and provide mechanical support for two other engine components, namely the exhaust valves and the fuel injectors.
- Camshaft(s): The camshaft is one of the most critical engine parts because it ensures timing of exhaust valves opening/closure, as well as fuel injection.
- Fuelling system: This is comprised of the high-pressure fuel pumps, the high-pressure pipelines and the fuel injectors (there can be more than one for each cylinder).

Diesel engines used for ship propulsion are tuned in order to operate near-steady-state on a well defined operating curve, the so called "fouled-hull" or propeller curve. The propeller loading curve, valid for full-bodied hulls, is of the form:

$$P = K_{pow} \cdot N^3 \Leftrightarrow \log P = \log K_{pow} + 3 \cdot \log N \tag{1.1}$$

where P is power in kW, N is shaft rpm and K_Q is the propeller law constant. In terms of torque Q the propeller curve (or law) takes the form:

$$Q = K_Q \cdot N^2 \Leftrightarrow \log Q = \log K_Q + 2 \cdot \log N \tag{1.2}$$

Note that the above cubic relationship for power represents propeller demands only in approximation, because it is dependent upon a variety of additional parameters for hull resistance and propulsion components; in effect, it assumes a more complex functional relationship. In practice, however, the cubic approximation is generally valid over limited power ranges.

Engine–propeller matching is done using a plot of the allowed engine operating envelope, where propeller power demands *vs.* shaft rpm are indicated, too, as in Figure 1.2. Major operational limit lines for large marine Diesel engines, shown below, are dictated by combustion efficiency (surge or smoke line) in the low rpm range and by shafting system bearings strength (torque limit line) in the higher rpm range.

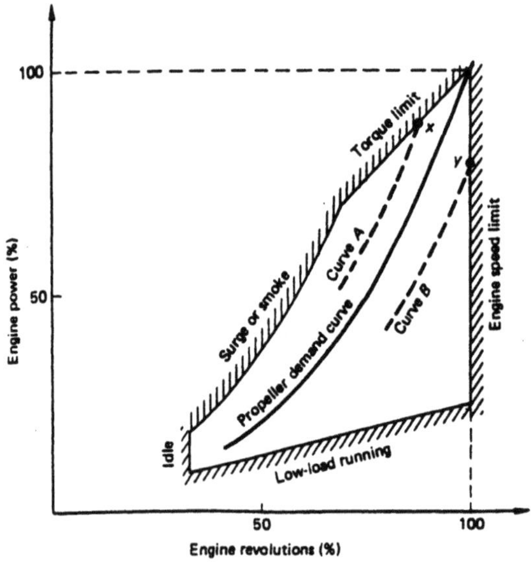

Figure 1.2 Chart used for engine-propeller matching

In the above plot, the propeller demand curve is shown to pass through the Maximum Continuous Rating (MCR) of the engine. The coordinates on the power-rpm chart of MCR represent the peak value of (continuous) engine power delivery and the engine speed limit.

Propeller power absorption characteristic can be modified during service due to a variety of factors such as sea conditions, wind strength, hull condition (roughness and fouling) and vessel displacement. Generally, increased severity of any of these factors requires a power increase in order to drive the ship at the same speed. In turn, this has an effect of moving the propeller power demand curve leftwards towards Curve A. Therefore, allowances need to be made for the propulsion plant to be able to develop full power under less favourable conditions due either to weather deterioration, deeper draught or hull fouling. In effect, a line like Curve B, located rightwards of the nominal (ideal) propeller demand curve, is selected for engine–propeller matching. This concept of difference in performance introduces the term "sea margin", in order to ensure that the ship propulsion plant has sufficient power available in service and throughout the docking cycle.

Steady-state engine load is expressed in terms of power rating, *i.e.* as a fraction of MCR power. However, the operating point of the engine is controlled through the position (in mm) of the fuel index (rack) at the fuel pump(s), which in most cases is provided in dimensionless form reduced to the interval [0,1] or [0%,100%]. The steady-state fuel index position is directly proportional to steady-state engine power, if the engine's mechanical losses are neglected. This implies that the fuel pumps of marine Diesel engines are designed in order to provide linear operation in terms of generated power. In fact, the engine power delivery is not linearly dependent on the fuel mass injected in the cylinders per cycle. The fuel mass required at each loading point on the propeller curve is calculated using a combination of theoretical thermodynamics, simulation results and testbed experimental data series. Then the mechanical design of the fuel pump is modified in order to achieve linear steady-state engine response in terms of power to fuel index changes.

Another important aspect of marine engine operation is that engine torque delivery is proportional to index position for constant engine speed (rpm) and provided that the engine is running with "excess air", *i.e.* adequate air supply for perfect combustion. Note that this comes is not contradiction with the linear steady-state power dependence upon index, as the rpm value is different from one steady-state operating point to another.

The engine mechanical power delivery is determined by the following design features:

- Number of cylinders (z_c)
- rpm at MCR (N_{MCR})
- Cylinder bore and piston stroke; these parameters determine the volume V_h swept by the piston displacement during a stroke (piston displacement volume), *i.e.*:

$$V_h = \frac{\pi \times (\text{bore})^2}{4} \times (\text{stroke}) \tag{1.3}$$

- Brake Mean Effective Pressure (BMEP, p_e), defined as follows:

$$p_e = \frac{P}{z_c \cdot V_h} \cdot \frac{1}{N/60} \tag{1.4}$$

  where P is power in watts and N is shaft rpm.

  Maximum BMEP is observed at MCR where $P = P_{MCR}$ (MCR power) and $N{=}N_{MCR}$. As argued in the next chapter, engine torque is directly proportional to BMEP, and therefore maximum engine torque delivery is observed at MCR as well.

- Indicated Mean Effective Pressure (IMEP, p_i) and Friction Mean Effective Pressure (FMEP, p_f). IMEP features the per-cycle average in-cylinder pressure, and FMEP the mechanical losses of the engine. The relationship between IMEP, FMEP and BMEP is:

$$p_e = p_i - p_f \tag{1.5}$$

- The maximum in-cylinder pressure value (maximum pressure, p_{max}) achieved in-cylinder per cycle, which affects the power output of the engine. At peak load, state-of-the-art large marine engines achieve p_{max} values above 130 bar.

Taking into account the fact that direct coupling of the propulsion engine to the propeller imposes an upper limit in the range 60–250rpm, propulsion power is increased by increasing V_h, BMEP or the number of cylinders. This remark has determined the trends in the evolution of the modern marine Diesel engine. Indeed, today's high-power-output engines are both long-stroke and very-large-bore in order to achieve maximisation of V_h. Piston stroke for these types of engine is typically larger than 2 m and cylinder bore above 900 mm (actually approaching 1 m).

BMEP of large marine engines has reached peak values of 18–18.5 bar. Limitations to further increase of BMEP originate from the fact that increasing BMEP leads inescapably to increasing in-cylinder p_{max}. Indeed, BMEP of 18 bar corresponds roughly to p_{max} values of about 130–140 bar. A great amount of research effort is nowadays invested in materials technology in order to manufacture combustion chambers with endurance to even higher values of p_{max}. This will enable a further increase of BMEP and engine power output.

Today's high-power two-stroke marine engines are most commonly built in arrangements of 6–12 cylinders according to requirements of propulsion power, engine room availability and mechanical vibration considerations. Indeed, more engine cylinders require larger engine rooms and impose modifications on the engine turbocharging system as the intake and exhaust manifolds grow longer. Also, a larger number of engine cylinders may be prohibitive due to limitations on crankshaft length and weight.

Finally, it is mentioned that today's marine propulsion engines are always turbocharged. For large two-stroke marine engines the constant pressure turbocharging system is used. With this type of turbocharging the exhaust ports of all cylinders are connected to a common receiver, whose volume is sufficiently large to damp the exhaust pressure pulses. Thus one or more turbochargers with a single turbine entry can be used. The main advantages of the constant pressure system are the simple exhaust receiver configuration and the almost steady conditions at the turbine inlet. The disadvantages are inadequate boost pressure at part engine loads and slow system response. Note that turbocharging introduces an intrinsic closed loop, as well as time delay and uncertainty to the system, and may, therefore, affect system relative stability and transient response. Specifically, the engine becomes more sensitive to torque demand fluctuations. Additionally, the linear relationship between engine torque and fuel index may cease to hold if the turbocharging system fails to deliver adequate air mass to the combustion chambers and, therefore, an incomplete combustion regime prevails.

1.1.3 The Screw Propeller

The single large-diameter propeller used for large cargo ship propulsion is a fixed-pitch mechanical construction with diameter that exceeds 5–6 m and composed of 3–6 blades. Screw propellers located aft-ship were introduced in ship propulsion not earlier than the 19th century. Their main advantage over other propulsion schemes is their superior hydrodynamic efficiency combined with simplicity of construction and operation. The efficiency of a propeller is defined as the ratio between the thrust power transmitted to the water and the mechanical rotational power provided to the propeller by the shaft, *i.e.*:

$$\eta_B = \frac{P_T}{P_D} \tag{1.6}$$

The overall torque Q_L developed by the water and exerted to the propeller shaft is given as follows:

$$Q_L = K_{q,prop} \cdot \rho \cdot D_{prop}^5 \cdot N^2 \tag{1.7}$$

where D_{prop} is the propeller diameter, ρ is the sea water density and $K_{q,prop}$ is a constant. Comparing this relationship with the one given for direct-drive engine load torque given earlier (Eq. (1.2) it is easily seen that:

$$K_Q = K_{q,prop} \cdot \rho \cdot D_{prop}^5 \tag{1.8}$$

Significant uncertainty is introduced to propeller torque, which forms the load torque applied to the propulsion engine, due to the fact that $K_{q,prop}$ is not a constant in the mathematical meaning of the word. Actually, coefficient $K_{q,prop}$ depends strongly, for fixed-pitch propellers, on the advance coefficient J_{screw}:

$$J_{screw} = \frac{v_{adv}}{N \cdot D_{prop}} \tag{1.9}$$

where v_{adv} is the advance speed (in m/s^2) of the propeller relative to the water, which is approximately equal to the advance speed of the ship. $K_{q,prop}$, and in effect K_Q, is a decreasing function of J_{screw}.

An rpm limitation is imposed on engine operation due to its direct coupling to the propeller as well. This limitation originates from the dependence of propeller efficiency η_B on coefficient J_{screw}. It can be seen that η_B has a peak value (maximum) when J_{screw} takes a specific value that lies somewhere between 0.3 and 0.7. As J_{screw} moves away from this specific value η_B degrades rapidly. This means that engine/propeller rpm must not exceed a comparatively small value (between 60 and 250 rpm), otherwise propeller efficiency becomes significantly poor (below 30%). Furthermore, $K_{q,prop}$ depends on the cavitation status of the propeller (quantified by the cavitation number σ_{cav}. In general, $K_{q,prop}$ increases with σ_{cav}. In turn, the cavitation number decreases as N (propeller rpm) and/or v_{adv} (propeller advance speed) increase.

Last, but not least, another important source of uncertainty is propeller inertia. When calculating the engine–propeller combined inertia, which determines the integration constant of the shaft, the propeller-entrained water inertia has to be taken into account, as well. However, the mass, and in consequence the moment of inertia, of the entrained water varies significantly. Expressed as a percentage of the propeller inertia, the entrained water can impose a surplus varying between 5 and 30%.

1.2 Contribution of this Work

1.2.1 Statement of the Problem

The subject of this text is robust control design methods for the marine Diesel propulsion system. Robust engine control should act in a manner ensuring safe powerplant operation, especially at near MCR and under conditions that may induce significant propeller load demand fluctuation. Such situations occur under a variety of vessel operating conditions, including mainly heavy weather and rough sea situations. Fluctuation of this magnitude has a number of undesirable effects on the powerplant operation, especially if it is run close to the upper bound of the allowed envelope. The most important consequence of such an event is the occurrence of critical main engine overspeed [4]. Indeed, a large propeller torque demand sink results in acceleration of the engine–propeller shaft. Then, if the engine rpm is set near MCR, the actual engine speed may exceed the maximum allowable limit, leading, thus, to critical overspeed and emergency shutdown, due to prohibitive main engine overloading. In order to avoid this undesirable situation, the main propulsion engine operating point is preventively reduced over the complete time interval for which heavy weather and rough sea conditions are experienced. As a result the ship speed decreases (voluntary speed loss).

As a consequence, it is required to increase the sea margin of the main propulsion engine in order not only to have reserves to overcome the inescapable hull fouling, but also to have the possibility to provide adequate propulsion power under rough sea conditions without the risk of an emergency engine shutdown due to critical overspeed. Therefore, the (resultantly larger) propulsion plant has increased installation costs and, possibly, increased running costs if the optimum operating point is positioned near MCR. Furthermore, even under fair sea states (*e.g.* sea state 3 or 4) it has been reported that significant propeller torque demand fluctuation can occur under certain conditions, *e.g.* when a large containership exhibits significant rolling due to beam sea encounter [4]. This can impose prohibitive limitations (enhanced fuel index limiter activity) to near MCR propulsion plant operation, and leading eventually to voluntary reduction of engine speed setpoint. This often results in failure in keeping the trading schedule, as it is another form of voluntary speed loss.

Limiters are incorporated in modern electronic control units for marine engines for protection against critical and off-design operation. Their effect is imposed directly on the control action, *i.e.* the fuel quantity injected per cycle in the engine cylinders, as expressed in dimensionless manner with the fuel index position percentage, and it is a non-linear and rather empirical feedback form of control. Specifically, a limiter is an upper or lower limit to fuel index position; the limit values are dependent upon the value of certain measured plant variables such as engine speed (rpm) or boost (scavenging) pressure. If the value of the control action generated by the linear part of the controller (usually a PI control law with appropriate gain values) is outside the range dictated by the limits, then it is saturated to one of the two bounds.

Closer investigation of this operating deficiency, later in the text, demonstrates that if the worst-case disturbance has been taken into account during the controller's linear part synthesis then the limiter activity could have been greatly reduced or even completely avoided, ensuring at the same time reliable plant operation. Moreover, it was made clear, in the framework of a wider research effort on marine control [4,5], that a more systematic methodology is required for marine plants with enhanced capabilities of control. The objective in this effort is the development of marine engine control systems that are robust against operating conditions different than the "nominal" calm sea ones. The conventional PI speed governors, used today in practice, are most commonly tuned for calm sea conditions.

1.2.2 Overview of the Approach

The approach towards the solution of the operational problems stated above is based on: (a) engine and propulsion plant modelling for control, (b) model linearisation based on reasonable assumptions and, finally, (c) designs for two proposed robust controllers, one with the PID control law and one with full state feedback. The specific methodological steps are outlined below.

Prior to control system development, understanding of the open-loop (uncontrolled) plant dynamics is needed. In standard control engineering practice this is done by formulating transfer function or state space models that quantify the transient response of the open-loop plant. For a number of reasons explained later in this text (including ship trading schedule, feasibility limitations, *etc.*), it is preferable to tune either the transfer function or the state-space model of the marine plant using physical (thermodynamic) engine simulation models, rather than performing shipboard measurement and experimentation campaigns. Therefore, in the context of this work, engine operation from the energetic/thermodynamic viewpoint is analysed for control purposes. Then, based on the insight acquired, a transfer function and a state-space model are established for the marine propulsion installation.

Using a reduced-order transfer function for the marine plant, an alternative PI(D) speed regulator tuning method is presented. The method relies on loop-shaping for meeting the disturbance rejection specification of the closed-loop transfer function. Although the method is exemplified for marine propulsion plants, it can be extended to any process that is described by a transfer function with a single, stable dominant pole. As D-term control is needed, an alternative is given for the case of marine propulsion plants to overcome the difficulties encountered in the practical implementation of signal differentiators. The PI(D) controller design method proposed is tested in the case of an actual propulsion plant of a large containership. A specific assessment of the PI(D) gains obtained using the proposed method is performed, confirming that they can provide adequate worst-case disturbance rejection, especially when compared to the PI governor used at the actual installation. Additionally, robustness against neglected dynamics is examined, using the full-order transfer function, which has been identified on the basis of thermodynamic engine simulation models.

The need for a systematic, formal approach to the design of the feedback propulsion controller is dealt with by firstly formulating a non-linear state-space model for the propulsion powerplant. This is achieved by combining the non-linear mapping abilities of neural nets with the extensive training (calibration) sets obtained using a cycle-mean, quasi-steady thermodynamic engine simulation model.

Finally, based on the plant state-space description, a full-state-feedback controller design methodology is proposed as an adaptation and application of the formal $H\infty$-synthesis and real parametric uncertainty analysis frameworks of robust control theory. In this technique, the propeller fluctuation is treated as a disturbance signal that has to be rejected by feedback control. Then, due to the additional real parametric uncertainties introduced to the state-space model by the fluctuation of the thermodynamic properties, as well as by the varying propeller-entrained water inertia, robustness analysis theoretical tools can be employed for the closed-loop system with state-feedback controls. In effect, a complete propulsion control system is proposed combining supervisory control, for smoother engine running, with feedback control for bounding the actual powerplant operation in close vicinity to the nominal desired behaviour.

1.2.3 Text Outline

A brief reference to the topics covered in the upcoming chapters is now given.

In Chapter 2, thermodynamic analysis of the turbocharged marine Diesel engine is performed. The physical processes of power torque generation are examined, resulting in a cycle-mean, quasi-steady model of engine operation that provides adequate insight, as well as a validation platform for control developments. The simulation model is exemplified in the case of a typical marine engine and the numerical solution procedure is explained and evaluated.

Chapter 3 deals with the problem of propulsion powerplant modelling for control purposes. The modelling starts with shafting system dynamical analysis, aiming to depict the effect of the engine–propeller shaft dynamics on controller design. Then the transfer function of the propulsion powerplant is formulated, using the "black-box" approach, in combination with *ad hoc* assumptions for the dynamics of the marine plant. Finally, identification is performed by employing a detailed, filling-and-emptying thermodynamic model of the engine processes. The procedure is validated using the propulsion powerplant of a large containership.

In Chapter 4, PI and PID control of the marine propulsion powerplant are examined from the viewpoint of modern linear robust control theory, and, specifically, $H\infty$ disturbance rejection (attenuation). Based on the analysis of Chapter 3, the closed-loop (with a PI(D) controller) scalar transfer function is formulated. Then, the PI(D) gains are calculated so that the $H\infty$-norm of the compensated system is equal or below specification. In effect, PI and PID compensated plants are assessed for robustness against neglected dynamics. From that perspective, superiority of PID over PI regulation is demonstrated. Finally, based upon the shafting system dynamical analysis of Chapter 3, an alternative method for implementing the D-term in practical installations is proposed. This method does not require differentiation of the rpm feedback signal, as the rpm derivative is calculated using the shaft torque feedback signal.

In Chapter 5, the state-space description of the marine plant is deduced from the thermodynamic engine model of Chapter 2. The analysis is based upon the neural net capabilities to depict non-linear mappings, if trained properly. State equations of the plant are then formulated, incorporating the neural torque approximators and the propeller law. The parametric uncertainty, present in the state equations, is in effect located and assessed. Next, the procedure for linearisation of the marine powerplant equations is applied. Finally, the open-loop transfer function matrix is determined and comparison to the scalar transfer function obtained in Chapter 3 is performed.

Chapter 6 deals with the marine propulsion powerplant control problem using state-feedback linear robust control theoretical results, in combination with open-loop optimised schedules for operating point changes. The disturbance rejection specifications are appropriately decomposed based on the analysis of Chapter 5. In effect, gains of the controller are calculated. Finally, applicability of criteria for robust stability and performance, as well as the effect of integral control on steady-state error, are briefly examined.

Chapter 7 concludes this work. Assessment of the modelling approaches, as well as of the PI(D) and state-feedback control options investigated, is done. Proposals are given for future research and investigations.

CHAPTER 2

MARINE ENGINE THERMODYNAMICS

2.1 Physical Engine Modelling

The significance of large two-stroke turbocharged Diesel engines for ship propulsion has been extensively analysed in the previous chapter. Indeed, the marine engine forms the most critical part of the propulsion powerplant of any modern cargo vessel. From a higher-level control point of view, the engine can be regarded as an actuator for ship propulsion. Because the engine is the propeller prime mover, any command for ship acceleration or deceleration has to be translated to an appropriate change of engine operating (loading) point. If for example, ship speed is reduced gradually due to developing rough head seas, engine loading has to be increased provided that MCR is not exceeded. In that respect, all other external factors affecting ship speed can be regarded as either disturbances, *e.g.* weather/sea conditions, or uncertainties, *e.g.* hull fouling status.

However, not only the ship speed–engine load relationship, but engine operation itself is non-linear and complex and subject to optimisation using appropriate control schemes. The engine control task also has to be reconsidered, accounting for potential advantages offered by state-of-the-art digital electronics and sensor/actuator technologies in combination with advanced control synthesis methods [4,5]. On the other hand, the common practice in advanced control design methods is to use state-space mathematical models for the plant, as well as for the controller. Whenever such plant model(s) are unavailable, physical principles, possibly incorporated in detailed physical simulation models, have to be used in order to deduce state-space mathematical models that can, in effect, be used for controller synthesis.

In the case of marine Diesel engines no state-space models are usually available. This is due to a number of reasons, the most important of which are the following:

- High costs do not allow construction of marine engines just for testing and conduction of experiments. In the automotive industry, on the other hand, construction of testbeds and test engines is a significantly easier task. Therefore, the conduct of experiments is a safe and realisable way to formulate the state variable maps depicting the system dynamics [6], *i.e.* the relationship of the state's temporal derivative and the state variable values. On the other hand, building of a large marine engine is not commenced until a specific order has been assigned to the manufacturer and the ship is in its initial stages of construction. Also, the construction of large marine engine testbeds with capabilities of transient loading is difficult and costly and, in effect, in many cases practically impossible.

- The engine physical/chemical processes are of high complexity and nonlinearity [2]. Although the differential equations of the engine and turbocharger shaft dynamics are similar to the ones encountered in any other electromechanical system, the physicochemical processes involved in combustion, *i.e.* the processes of power/torque generation, are highly non-linear and complex. Indeed, the equations governing these phenomena are more similar to the ones encountered in chemical plants and process industry. Therefore, a "simple" linearised approach cannot be used in a straightforward manner and without taking into account the limitations and approximations involved.

Due to the lack of extensive experimental data series, especially for transient operation a variety of analytical thermodynamic and Computational Fluid Dynamics (CFD) models and methodologies exist for performance prediction of large marine engines. These models rely on the physical principles of thermodynamics, fluid dynamics and chemical kinetics in order to predict steady-state and transient performance of the engine prior to its manufacture. Their accuracy varies, but in general even the simpler ones can provide adequate accuracy for control development purposes. Therefore, for the present and the foreseeable future, using such physical simulation models is a major practical and reliable way to obtain significant insight to the engine physicochemical processes.

A very important class of such physical engine models is the so-called quasi-steady, cycle-mean-value thermodynamic models. Quasi-steady models are based on the assumption that the process equations that are used for steady-state analysis can be extended appropriately to dynamic/transient situations. The objective is to estimate the engine-cycle-averaged (from cycle to cycle) temporal evolution of thermodynamic (pressures and temperatures) and mechanical variables of interest. Furthermore, the engine is a highly spatially distributed system, especially when the thermodynamic variables are considered. Quasi-steady models are based on reasonable assumptions in order to eliminate the distributed character of thermodynamic variables and deduce spatial means for each one of these variables [7,8]. This includes the appropriate partitioning of the turbocharger/engine interconnected volumes (plenums) forming thus a lumped-parameter (zero-dimension) model. In conclusion, a set of intermediate variables is formulated, which are the cycle/plenum means of their counterpart thermodynamic distributed variables.

The mathematical relations between these intermediate variables constitute a perplexed non-linear algebraic system of equations, which, if solved, can provide the engine, turbine and compressor torques as functions of engine and turbocharger shaft rpm (state variables), as well as fuel index position (control action). Then, the differential equations of motion (Newton's law of motion) for the two shafts (engine and turbocharger) of the powerplant can be solved.

A typical quasi-steady cycle-mean-value thermodynamic engine model is presented in more detail in the rest of the chapter, while methods to overcome the numerical solution of the algebraic system involved, are presented in Chapters 3 and 5.

2.2 Turbocharged Engine Model Variables

The major parts of a typical, large turbocharged marine Diesel engine are described in Table 2.1 along with the lumped/cycle-averaged thermodynamic variables of interest "attached" to each one of them.

Table 2.1 Engine thermodynamic variables of interest

Turbocharger compressor	Air mass flow rate ($\dot{m}_A$)
Intercooler	Compressed air temperature (T_C) Intercooler efficiency (η_{IC})
Scavenging receiver	Scavenging pressure (p_I) Scavenging temperature (T_I) Scavenging air enthalpy (h_I)
Engine cylinders	Fuel mass flow rate ($\dot{m}_F$) Air-to-fuel ratio (A/F) Combustion efficiency (η_c) Brake Mean Effective Pressure (BMEP, p_e) Indicated Mean Effective Pressure (IMEP, p_i) Friction Mean Effective Pressure (FMEP, p_f) Fuel chemical energy proportion in exhaust gas (ζ_a)
Exhaust receiver	Exhaust pressure (p_E) Exhaust temperature (T_E) Exhaust gas enthalpy (h_E)
Turbocharger turbine	Exhaust mass flow rate ($\dot{m}_E$) Turbine flow coefficient (a_T) Turbine isentropic efficiency (η_{iT})

The above intermediate variables can be calculated if the powerplant operating point is known. The plant operating point is, in turn, determined if:

- Engine crankshaft rotational speed (rpm) N_E,
- Turbocharger shaft rotational speed (rpm) N_{TC} and
- Fuel index (rack) position F_R

are given. Additionally, the following external variables are also necessary for the determination of the engine operating point:

- Ambient (atmospheric) pressure (p_a, typical value 1 bar = 10^5 N/m^2)
- Ambient (atmospheric) temperature (T_a, typical value 290 K = 17°C)
- Intercooler coolant (water) temperature (T_w)

However, these variables are treated rather as fixed-value model parameters than as variables. Indeed, the ambient conditions (p_a and T_a) are maintained approximately constant in any modern ship's engine room by use of ventilation and/or air conditioning systems, in order to sustain machinery performance (including the main engine) at specification levels. For the same reason the coolant temperature is maintained within a narrow range using an external heat exchanger, possibly with controllable heat transfer capacity.

The physical principle, employed for the quasi-steady, cycle-mean-value modelling of the turbocharged two-stroke Diesel engine, is its thermodynamic equivalence of the air and exhaust gas flows through orifices. In that respect, modelling of the air flow through the engine is considered equivalent to that through an orifice with effective area $\bar{A}_{Veq}$ remaining constant over time. Parameter $\bar{A}_{Veq}$ is determined by the geometrical configuration of the exhaust valves and the inlet ports of the cylinders, as well as the valve timing and the number of cylinders. It is assumed to remain constant over time, in spite of the interrupting communication of the inlet to exhaust ports, because it is calculated as a mean value over a complete engine cycle. Note that in the case of two-stroke engines the thermodynamic cycle coincides with one crankshaft revolution. In the case of four-stroke engines this is not the case; one thermodynamic cycle requires two revolutions. The turbocharger turbine is also considered as an orifice through which the exhaust gas is flowing. The turbocharger compressor is considered as a pump delivering pressurised air to the engine. Its output pressure, as well as the load torque imposed on the turbine through the turbocharger shaft, are functions of the turbocharger rpm. The picture is completed by the temperature/mass increase due to fuel injection and combustion, as well as the air temperature drop at the intercooler.

The above physical modelling approach, of the one pump (compressor) and two orifices (engine ports and valves, as well as turbine) connected in series is shown in Figure 2.1, where the various gas flows are depicted, too.

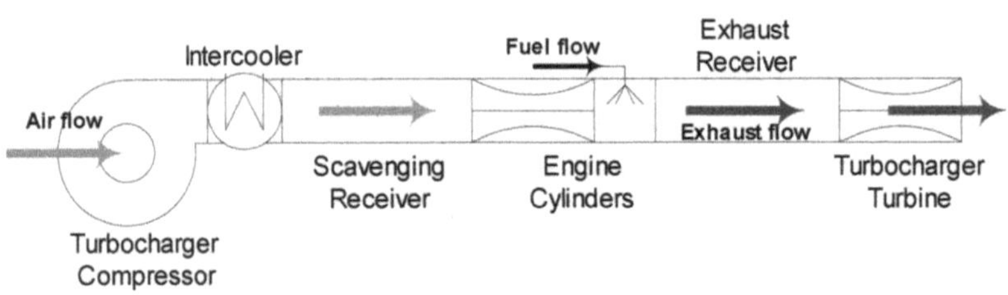

Figure 2.1 Thermodynamic modelling approach

The objective of the thermodynamic analysis, however, remains the calculation of engine, turbine and compressor torques that govern the propulsion system dynamics, through the differential equations presented in the next section.

2.3 Turbocharged Engine Dynamical Equations

The propulsion system dynamics are governed by the engine–propeller and turbocharger shaft dynamical equations. In the present analysis no damping is assumed for the two shafts, *i.e.* they are both considered to be fully elastic. The equations are then simplified significantly without major loss in accuracy.

For crankshaft rpm (N_E) or propeller rpm (N) the differential equation is as follows:

$$\dot{N}_E(t) = \frac{Q_E - Q_L}{I_{total}} \tag{2.1}$$

where Q_E is the cycle-mean torque delivery of the engine, Q_L is the cycle-mean torque demand (load) imposed on the powerplant's shafting system and I_{total} is the total shafting system inertia averaged over a cycle.

Under the assumption of engine–propeller direct coupling (no gearbox or clutch):

- No reduction ratio is introduced between engine and propeller rpm, *i.e.*:

$$N \equiv N_E \tag{2.2}$$

- Propeller-law engine loading can be considered, *i.e.*:

$$Q_L(N_E) = K_Q \cdot N_E^2 = (K_{Q0} + \Delta K_Q) \cdot N_E^2 \tag{2.3}$$

where K_Q is the propeller torque coefficient, on which some uncertainty k_Q has been superimposed on the nominal value K_{Q0}, according to the considerations presented in Chapter 1

- The shafting system inertia can be expressed as follows:

$$I = I_E + 1.15 \cdot I_{prop} + \Delta I = I_0 + \Delta I \tag{2.4}$$

i.e. as sum of the engine crankshaft inertia I_E plus the propeller inertia I_{prop} augmented by a (nominal) 15% accounting for the entrained water inertia, as analysed in Chapter 1. Note that some uncertainty ΔI is also superimposed on the nominal value $I_0 = I_E + 1.15 \cdot I_{prop}$ of this parameter, too, due to the variation of the entrained water inertia.

Engine torque delivery Q_E is calculated as a function of the thermodynamic variables. Specifically, it is argued that engine torque is directly proportional to BMEP of the engine. In Chapter 1, BMEP has been defined as the in-cylinder pressure value to which the engine rotational power delivery P corresponds:

$$P = p_e \cdot z_c \cdot V_h \cdot \frac{N}{60} \tag{2.5}$$

However, the following holds for mechanical power:

$$P = Q_E \cdot \frac{2\pi}{60} N \tag{2.6}$$

By equating the above:

$$Q_E = \frac{z_c \cdot V_h}{2\pi} \cdot p_e \tag{2.7}$$

Also, it should be mentioned that some dead time τ is involved with engine torque generation after a change of fuel index position F_R. This is due to the fact that torque generation is effectuated after fuel has been injected in the engine cylinders and combustion has taken place. Therefore, the new torque value, dictated by the modified fuel index position, has been fully achieved after one complete crankshaft revolution in two-stroke engines (or two revolutions for four-stroke engines). In the meantime, torque is changing by a manner dictated by the firing order (sequence) of the engine cylinders. The engine torque dead time lies in the following range, for two-stroke engines:

$$\frac{1}{4 \cdot N_E} < \tau < \frac{1}{4 \cdot N_E} + \frac{1}{z_c \cdot N_E}, \ \tau \text{ in min} \tag{2.8}$$

However, this effect (delay) becomes significant only for low engine speeds, typically below 1 rps = 60 rpm). Furthermore, in the worst case:

$$\tau = \frac{1}{4 \cdot N_E} + \frac{1}{z_c \cdot N_E} < \frac{1}{N_E}, \text{ for any } z_c \geq 2 \tag{2.9}$$

In conclusion, engine torque delay (τ) effect can be neglected in cycle-mean-value engine models, because this delay is smaller than the sampling time interval (step) of this model. The time step is the period of one crankshaft revolution, *i.e.* (in min) $1/N_E$.

The turbocharger dynamics are depicted in the following differential equation, which is analogous to the one holding for the engine–propeller shaft:

$$\dot{N}_{TC}(t) = \frac{Q_T + Q_C}{I_{TC}} \tag{2.10}$$

where Q_T is the torque delivery of the turbine and Q_C is the load torque of the compressor. I_{TC} is the combined moment of inertia of the turbocharger that includes the inertia of the turbocharger shaft, as well as the inertias of the turbine wheel and compressor impeller. No uncertainty is included in this system parameter, as the inertias involved do not change.

Q_T and Q_C are calculated from thermodynamic variables with the following algebraic relations:

$$Q_T = \frac{\eta_{iT} \cdot C_{\mathrm{P,exh}} \cdot T_E \cdot \dot{m}_E}{{}^{2\pi}\!/_{60} \cdot N_{TC}} \cdot \left[1 - \left(\frac{p_a}{p_E}\right)^{\frac{\gamma_E - 1}{\gamma_E}}\right] \tag{2.11}$$

$$Q_C = \frac{C_{\mathrm{P,air}} \cdot T_a \cdot \dot{m}_A}{\eta_{TC} \cdot {}^{2\pi}\!/_{60} \cdot N_{TC}} \cdot \left[1 - \left(\frac{p_I}{p_a}\right)^{\frac{\gamma_A - 1}{\gamma_A}}\right] \tag{2.12}$$

Constants γ_A and γ_E are the specific heat ratios for air and exhaust gas respectively:

$$\gamma_A = 1.4, \quad \gamma_E = 1.34 \tag{2.13}$$

Note that the above parameters are dependent upon air and exhaust gas temperature respectively. However, the above values are typical for the usual air temperature (288 K or 15°C) and exhaust gas temperature (400–1100 K).

$C_{\mathrm{P,air}}$ and $C_{\mathrm{P,exh}}$ are the specific heat at constant pressure of air and exhaust gas respectively. In general, C_{P} is calculated according to the following mathematical relationship:

$$C_{\mathrm{P}} = \frac{\gamma - 1}{\gamma} \cdot \frac{R}{\mathrm{M}_{mol}} \tag{2.14}$$

where R is the ideal gas constant $\left(R = 8.314 \frac{\mathrm{J}}{\mathrm{mol \cdot K}} \right)$ and M_{mol} is the molecular weight of the gas considered. In the case of air and exhaust gas it is assumed that:

$$\mathrm{M}_{mol,\mathrm{air}} = 28.96 \text{ gr/mol}, \quad \mathrm{M}_{mol,\mathrm{exh}} = 30.0 \text{ gr/mol} \tag{2.15}$$

By substituting the values of γ and M_{mol} in the general formula one obtains:

$$C_{\mathrm{P,air}} = 1005.0 \frac{\mathrm{J}}{\mathrm{kg \cdot K}}, \; C_{\mathrm{P,exh}} = 1117.0 \frac{\mathrm{J}}{\mathrm{kg \cdot K}} \tag{2.16}$$

Constant η_{TC} finally includes the mechanical efficiency η_{mTC} of the turbocharger and the compressor isentropic efficiency η_{iC}. Both these parameters are assumed to be constant and close to unity. Specifically:

$$\eta_{TC} = \eta_{mTC} \cdot \eta_{iC} \overset{\eta_{mTC} \approx 0.99}{\Rightarrow} \eta_{TC} \approx \eta_{iC} \tag{2.17}$$

One can identify significant similarities between the relationships for turbine and compressor torques. This is why they are derived considering adiabatic air compression (at the compressor) and adiabatic exhaust gas expansion (at the turbine). Based on these theoretical thermodynamic processes the relationships for turbine and compressor torques can then be derived.

Note also that:

$$Q_T > 0 \text{ and } Q_C < 0 \tag{2.18}$$

This is because in the case of the compressor $p_I > p_a$ and in the case of the turbine $p_E > p_a$, while in both cases $(\gamma - 1)/\gamma > 0$.

For quasi-steady engine models the only dynamical equations are those concerning engine and turbocharger rpm. In other, more analytical thermodynamic engine models the gaseous mass accumulation in-cycle is also of interest. Therefore, differential relations for the air and exhaust flow rates have to be included, increasing the number of dynamical equations. However, for engine control purposes the cycle-mean-value approach suffices, provided that adequate modelling accuracy is guaranteed, because the control action (fuel index position) has an effect on engine operation once per cycle. In the following sections the algebraic interdependence of the thermodynamic variables, presented in the previous section and required for the calculation of engine, turbine and compressor torques, is analysed. The sequence of presentation tracks the gaseous flow through each one of the major parts of the powerplant.

2.4 Turbocharged Engine Algebraic Equations

2.4.1 Turbocharger Compressor

The flow of air through the turbocharger compressor is modelled as an adiabatic compression process. The compressor ensures the required air mass flow rate in order to sustain the perfect combustion regime in the combustion chambers. This regime maximises both engine efficiency and power/torque output. The air mass flow rate is considered in quasi-steady engine models to be continuous, as already mentioned. Restriction of air flow is imposed due to the presence of inlet ports and exhaust valves, which induce interruptions to continuous air flow. For analysis purposes it is assumed that air flows from the scavenging to exhaust receiver through an orifice of equivalent effective area $\bar{A}_{V_{eq}}$, which will be calculated, and pressure difference (drop) $(p_I - p_E)$. In effect:

$$\dot{m}_A = C_V \cdot \bar{A}_{V_{eq}} \cdot \frac{p_I}{\sqrt{R_{\text{air}} \cdot T_I}} \cdot \left\{ 2 \cdot \left(\frac{\gamma_A}{\gamma_A - 1} \right) \left[\left(\frac{p_E}{p_I} \right)^{\frac{2}{\gamma_A}} - \left(\frac{p_E}{p_I} \right)^{\frac{\gamma_A + 1}{\gamma_A}} \right] \right\}^{1/2} \tag{2.19}$$

where $R_{\text{air}} = \dfrac{R}{M_{mol,\text{air}}} = 287.056 \dfrac{\text{J}}{\text{kg} \cdot \text{K}}$.

Coefficient C_V is the resistance coefficient for air flow through the mean effective area of the inlet ports and exhaust valves. For two-stroke engines, due to the employment of inlet ports, instead of valves, C_V can be assumed equal to 0.9.

Equivalent effective area is calculated as the cycle mean of $A_{V_{eq}}(\omega)$, where ω is the crank angle. *I.e.*:

$$\bar{A}_{V_{eq}} = \frac{z_c}{2\pi} \cdot \int_0^{2\pi} A_{V_{eq}}(\omega) d\omega \tag{2.20}$$

Finally, $A_{V_{eq}}(\omega)$ can be calculated from the instantaneous (function of ω as well) openings of the engine inlet ports and exhaust valves, *i.e.*:

$$A_{V_{eq}}(\omega) = \frac{A_{inlet}(\omega) \cdot A_{exhaust}(\omega)}{\sqrt{A_{inlet}^2(\omega) + A_{exhaust}^2(\omega)}} \tag{2.21}$$

$A_{inlet}(\omega)$ and $A_{exhaust}(\omega)$ are engine configuration data provided by the engine manufacturer.

2.4.2 Intercooler

Compression of air causes a temperature rise in the scavenging air, given by the thermodynamic relation concerning adiabatic processes, *i.e.*:

$$T_C = T_a \cdot \left[\frac{\left(p_I / p_a \right)^{\frac{\gamma_A - 1}{\gamma_A}} - 1}{\eta_{iC}} + 1 \right] \tag{2.22}$$

This fact moderates the effect (increase) of compression on density. In order to recover this negative effect an additional air cooler (intercooler) is placed between the compressor outlet and the scavenging receiver. This is a water-cooled heat exchanger that can lower the temperature of scavenging air by transmitting part of the air's heat to the coolant. The efficiency of the intercooler is a decreasing function of air mass flow rate. As a first approximation a relationship of the following form can be considered for the intercooler efficiency η_{IC} :

$$\eta_{\mathrm{IC}} = 1 - \kappa_{\mathrm{IC}} \cdot \dot{m}_A \tag{2.23}$$

where κ_{IC} is a constant that can be calculated from manufacturer specification data sheets. Many authors assume that intercooler efficiency is approximately constant with a value above 95%. Note, also, that a pressure drop in scavenging air pressure is involved due to the intercooler presence. However, it is too low to affect modelling accuracy and, therefore, may be neglected.

2.4.3 Scavenging Receiver

The most important thermodynamic operating variables of an engine's scavenging air receiver are scavenging air temperature and pressure. Air pressure value is governed mainly by the turbocharger operating status, which in the case of quasi-steady models is assumed to be determined fully by its rotational speed (turbocharger rpm, N_{TC}). This assumption is valid, especially for marine Diesel plants, because the engine–turbocharger matching is optimised in order for these two major components to operate on a well-specified operating line (curve). This operating line is most commonly plotted on a compressor map chart. Coordinates of the map are corrected mass flow rate (x-axis), *i.e.* $\frac{\dot{m}_A \cdot \sqrt{T_a}}{p_a}$, and pressure ratio ($y$-axis), *i.e.* $\frac{p_I}{p_a}$. On such a map the constant corrected speed $\left(\frac{N_{TC}}{\sqrt{T_a}} \right)$ lines and constant compressor (isentropic) efficiency (η_{iC}) curves are also indicated. The intersection points of these lines define the various operating points of the compressor, which are located on the operating line dictated by engine–turbocharger matching requirements. For large marine engines matching is optimised in order for the compressor to operate approximately on a constant efficiency curve. This is why η_{iC} can be considered to remain constant for all operating points of interest. This assumption is valid even for fast transients because under such conditions the operating point moves in parallel with the selected constant η_{iC} curve and the deviations do not exceed 1–2%. However, the assumption ceases to hold with adequate accuracy if some kind of malfunction occurs.

Furthermore, a constant η_{iC} curve can be approximated by a second-order polynomial of corrected speed. This is argued if the enthalpy rise due to compression is expressed as function of turbocharger rpm, on the one hand, and of pressure ratio, on the other hand. Then the following relationship for pressure ratio can be deduced:

$$\frac{p_I}{p_a} = \kappa_C \cdot \left(\frac{N_{TC}}{\sqrt{T_a}} \right)^2 + 1 \tag{2.24}$$

Coefficient κ_C can be calculated if one point of the operating line is known, *i.e.* a pair of values for p_I and N_{TC}.

Another important thermodynamic variable, as already mentioned, is scavenging air temperature T_I. This is calculated as a function of the compressed air temperature T_C (which is higher than T_a due to compression) if the effect of the intercooler is taken into account:

$$T_I = (1-\eta_{IC}) \cdot T_C + \eta_{IC} \cdot T_w \tag{2.25}$$

where T_w is the cooling water temperature.

Finally, scavenging air enthalpy h_I is defined as:

$$h_I \equiv C_{P,air} \cdot T_I \tag{2.26}$$

2.4.4 Engine Cylinders

The two most important processes of engine mechanical power/torque generation take place in the engine cylinders:

- Fuel injection
- Combustion

The fuel pumps that are driven by the camshaft in large marine engines sustain fuel injection. Timing of fuel injection is managed by camshaft geometry. Lately, some Variable Injection Timing (VIT) electromechanical systems for marine engines have been developed and installed shipboard. Also, a new generation of marine engines with fully electronic injection timing (hydraulically actuated) is under development. However, quasi-steady models cannot depict the fuel injection timing effect, as they are concerned with cycle-mean values, rather than in-cycle evolution. Therefore, fuel injection timing is modelled only as an influence (increase or decrease) on power/torque values achieved with the same amount of fuel, *i.e.* as an increase or decrease of specific fuel consumption (SFOC, g/kWh). On the other hand, the fuel mass injected in the cylinders per cycle, expressed as fuel mass flow rate $\dot{m}_F$ (kg/s), is a very important variable. The $\dot{m}_F$ value depends on the fuel index (fuel rack, F_R) position as explained in Chapter 1. Also, it depends on engine rpm, according to the following:

$$\dot{m}_F = z_c \cdot \mathrm{m}_{\mathrm{F,max}} \cdot F_R \cdot N_E \tag{2.27}$$

where $\mathrm{m}_{\mathrm{F,max}}$ is the maximum amount (kg) of fuel that can be injected in one cylinder per cycle. Note that fuel index position F_R is expressed in dimensionless form and lies within the interval [0,1].

Fuel index also determines IMEP. Indeed, increasing the fuel index leads to higher IMEP, provided that a perfect combustion regime is maintained. In quasi-steady engine models the combustion regime can be quantified using air-to-fuel ratio A/F, which is defined as:

$$(\mathrm{A/F}) \equiv \frac{\dot{m}_A}{\dot{m}_F} \tag{2.28}$$

A/F in turn determines combustion efficiency η_c according to the following relationship:

$$\eta_c = \begin{cases} 1 & \text{, if } (\mathrm{A/F}) \geq (\mathrm{A/F}) \\ \dfrac{(\mathrm{A/F}) - (\mathrm{A/F})_{\mathrm{low}}}{(\mathrm{A/F})_{\mathrm{high}} - (\mathrm{A/F})_{\mathrm{low}}} & \text{, if } (\mathrm{A/F})_{\mathrm{low}} < (\mathrm{A/F}) < (\mathrm{A/F})_{\mathrm{high}} \\ 0 & \text{, if } (\mathrm{A/F}) \leq (\mathrm{A/F})_{\mathrm{low}} \end{cases} \tag{2.29}$$

IMEP is then calculated according to the following relationship:

$$p_i = \eta_c \cdot \mathrm{p}_{\mathrm{i,max}} \cdot F_R \tag{2.30}$$

Constant $\mathrm{p}_{\mathrm{i,max}}$ is the maximum IMEP that the engine can achieve and is specified by the manufacturer. The linear dependence of IMEP on fuel index is clear from the above relation, although there also exists a negligible dependence of IMEP on engine rpm. However, this dependence ceases to be linear if combustion efficiency degrades from unity. In turn, this can happen if A/F drops under $(\mathrm{A/F})_{\mathrm{high}}$ (typical value is 20–27 for HFO and 17–20 for Diesel oil). In the case that A/F is lower than the above limits combustion is not perfect and, therefore, the generated power/torque degrades. Linear degradation of IMEP, and in effect of power and torque, with A/F is assumed because a low value of A/F (lower than stoichiometric A/F, which lies in the range 14.5–15) means inadequate air mass for perfect fuel combustion. Therefore, it is assumed that the burnt mass of fuel is directly proportional to the available combustion air mass. However, it is also possible that combustion is not effectuated at all if A/F drops under a second limit $(A/F)_{low}$ (and not until zero) which is in the range 5–8. This is due to excessive air in-cylinder cooling, caused by fuel injection. The temperature of the mixture is then not high enough to start the combustion process and, therefore, IMEP and engine power/torque are driven to zero.

As already mentioned, BMEP, which is related proportionally to engine torque, is connected with IMEP and FMEP by the relation:

$$p_e = p_i - p_f \tag{2.31}$$

FMEP is usually calculated as a multi-linear (affine) function of IMEP and engine rpm, *i.e.*:

$$p_f = \kappa_{f2} \cdot p_i + \kappa_{f1} \cdot N_E + \kappa_{f0} \tag{2.32}$$

Constants κ_{f2}, κ_{f1} and κ_{f0} are provided by the engine manufacturer. Alternatively, for large marine propulsion Diesel engines, MAN-B&W specifies FMEP as a linear function of fuel index as follows [12,32]:

$$p_f = \kappa_{f3} \cdot F_R + \kappa_{f4} \tag{2.33}$$

However, this form of expressing FMEP can be transformed to the previous one (with κ_{f1} set to zero) if constants κ_{f3} and κ_{f4} are substituted as follows:

$$\kappa_{f3} = \kappa_{f2} \cdot \mathrm{p}_{i,max}, \; \kappa_{f4} = \kappa_{f0} \tag{2.34}$$

Also, note that combustion efficiency is assumed to be equal to unity.

The MAN-B&W FMEP form is more convenient for calculating BMEP in steady-state as a linear function of fuel index position:

$$p_e = p_i - p_f = \mathrm{p}_{i,max} \cdot F_R - \kappa_{f3} \cdot F_R - \kappa_{f0} = (\mathrm{p}_{i,max} - \kappa_{f3}) \cdot F_R - \kappa_{f0} \tag{2.35}$$

One is reminded that, in steady-state, the combustion efficiency is guaranteed to be unity, because the turbocharger matching ensures that adequate combustion air mass is delivered to the cylinders (*i.e.* $(A/F) > (A/F)_{high}$). Constant κ_{f0} is the pressure-equivalent of mechanical power losses in idle (or motoring) engine operating mode.

The last thermodynamic variable of interest in the cylinders is the fuel chemical energy proportion in exhaust gas, ζ_α. This parameter is correlated with BMEP using a linear function of BMEP. The functional dependence given below is valid for any two-stroke Diesel engine:

$$\zeta_a = \kappa_{Z1} \cdot p_e + \kappa_{Z0} \tag{2.36}$$

For two-stroke engines and BMEP expressed in (N/m^2): $\kappa_{Z1} = 0.0105 \times 10^{-5}$ and $\kappa_{Z0} = 0.3120$.

2.4.5 Exhaust Receiver

Exhaust pressure and temperature are very important because they can be used for benchmarking any zero-dimensional engine model as:

(a) they are relatively easily measured in either testbed or installed plants and engines;

(b) they comprise a measure for combustion modelling accuracy as the exhaust gas is the direct outcome of combustion;

(c) they determine turbocharger rpm, as the exhaust gas is flowing through the turbocharger turbine causing it to accelerate or decelerate according to its thermal properties in comparison with the environment.

Exhaust pressure is calculated using the mathematical relation of the exhaust mass flow rate through the turbine:

$$\dot{m}_E = \mathrm{a_T} \cdot C_\mathrm{T} \cdot A_{\mathrm{T}eq} \cdot \frac{p_E}{\sqrt{R_\mathrm{exh} \cdot T_E}} \cdot \left\{ 2 \cdot \left(\frac{\gamma_E}{\gamma_E - 1} \right) \cdot \left[\left(\frac{p_a}{p_E} \right)^{\frac{2}{\gamma_E}} - \left(\frac{p_a}{p_E} \right)^{\frac{\gamma_E + 1}{\gamma_E}} \right] \right\}^{1/2} \tag{2.37}$$

where $R_\mathrm{exh} = \dfrac{R}{\mathrm{M}_{mol,\mathrm{exh}}} = 277.133 \dfrac{\mathrm{J}}{\mathrm{kg} \cdot \mathrm{K}}$. The similarity of the above equation to the one used for calculation of air mass flow rate is obvious, as in both cases equivalence of flow through an orifice is assumed. Therefore, coefficient C_T is the resistance coefficient for exhaust gas flow through the mean effective area of the turbocharger turbine. Due to the advanced design of the turbine nozzle and wheel (rotor) blades for modern marine engine turbochargers it can be assumed that:

$$C_\mathrm{T} \approx 1.0 \tag{2.38}$$

In the same sense, $A_{\mathrm{T}eq}$ is the equivalent effective area of the turbine. This is calculated from the geometrical configuration data of the turbine. Note that due to the large number of blades on the rotor no cycle-averaging is needed as the effective area does not vary greatly during one engine cycle. The turbine effective area is calculated as follows:

$$A_{\mathrm{T}eq} = \frac{A_{nozzle} \cdot A_{rotor}}{\sqrt{A_{nozzle}^2 + A_{rotor}^2}} \tag{2.39}$$

where A_{nozzle} is the minimum flow area of the nozzle and A_{rotor} is the minimum flow area of the wheel. However, due to the complexity of the phenomenon calculated, $A_{\mathrm{T}eq}$ does not provide adequate modelling accuracy. In effect, the multiplicative flow correction parameter $\mathrm{a_T}$ is introduced in order to compensate the modelling accuracy error.

In the case of the turbine, choked flow may occur also if the value of exhaust pressure exceeds the threshold value dictated by the following inequality:

$$\frac{p_E}{p_a} \geq \left(\frac{\gamma_E+1}{2}\right)^{\frac{\gamma_E}{\gamma_E-1}} \approx 1.85 \tag{2.40}$$

If the above value is exceeded then the exhaust mass flow rate dependence on the turbine pressure ratio p_a / p_E ceases to hold and p_E is calculated by the simplified relation with $\dot{m}_E$:

$$\dot{m}_E = \mathrm{a_T} \cdot C_\mathrm{T} \cdot A_{\mathrm{T}eq} \cdot \frac{p_E}{\sqrt{R_\mathrm{exh} \cdot T_E}} \cdot \left\{ \gamma_E \cdot \left(\frac{2}{\gamma_E+1}\right)^{\frac{\gamma_E+1}{\gamma_E-1}} \right\}^{1/2} \tag{2.41}$$

Exhaust temperature T_E is calculated using the exhaust gas enthalpy h_E. This variable is dependent on the fuel's specific calorific value $\mathrm{H_U}$ (referred to 0 K), parameter ζ_a (fuel chemical energy proportion in exhaust gas), as well as A/F. Specifically:

$$h_E = \eta_\mathrm{exh} \cdot \frac{h_I + \frac{\zeta_a \cdot \mathrm{H_U}}{(\mathrm{A/F})}}{1 + \frac{1}{(\mathrm{A/F})}} \tag{2.42}$$

Coefficient η_exh stands for the exhaust temperature correction factor. Indeed, the quantity $\left(h_I + \frac{\zeta_a \cdot \mathrm{H_U}}{(\mathrm{A/F})}\right) \Big/ \left(1 + \frac{1}{(\mathrm{A/F})}\right)$ stands for the exhaust gas enthalpy at the entry point of the exhaust receiver and not at the entry point of the turbine, as it is calculated as an enthalpy increase to the scavenging air enthalpy h_I. This enthalpy increase is due to fuel injection and combustion and takes place in the engine cylinders. However, some enthalpy decrease takes place in the exhaust receiver due to the heat exchange process there. This is why there is a need for the introduction of η_exh, as the exhaust pressure and temperature values required for the calculation of exhaust flow rate $\dot{m}_E$ through the turbine are the ones at the turbine inlet and not those at the exhaust receiver entry. However, as a first approximation it can be assumed that $\eta_\mathrm{exh} \approx 1.0$, *i.e.* the effect of the heat exchange process at the exhaust receiver can be neglected.

Finally, the exhaust temperature T_E can be calculated directly from the exhaust gas enthalpy:

$$T_E = \frac{h_E}{C_\mathrm{P,exh}} \tag{2.43}$$

Remember that $C_\mathrm{P,exh} = 1117.0 \frac{\mathrm{J}}{\mathrm{kg \cdot K}}$.

2.4.6 Turbocharger Turbine

The exhaust mass flow rate is calculated as the sum (conservation of mass) of air mass flow rate $\dot{m}_A$ and fuel mass flow rate $\dot{m}_F$, *i.e.*:

$$\dot{m}_E = \dot{m}_A + \dot{m}_F \tag{2.44}$$

The behaviour of the turbine is commonly described by charts of swallowing capacity $\dot{m}_E$ through the turbine as a function of pressure ratio p_a / p_E. However, for modelling purposes the relationship presented in the previous section is adopted. The turbine behaviour is completed with the plots for the isentropic turbine efficiency η_{iT}. In general:

$$\eta_{iT} = \eta_{iT}\left(\frac{p_a}{p_E}, \frac{U_{\mathrm{T}}}{C_{\mathrm{S}}}\right) \tag{2.45}$$

where U_{T} is the velocity (in m/s) of the rotor blade tip:

$$U_{\mathrm{T}} = \pi \cdot D_{turb} \cdot N_{TC} \tag{2.46}$$

where D_{turb} is the turbine wheel diameter and C_{S} is the exhaust gas velocity. For isentropic processes:

$$C_{\mathrm{S}} = \sqrt{2 \cdot h_E \cdot \left[1 - \left(\frac{p_a}{p_E}\right)^{\frac{\gamma_E - 1}{\gamma_E}}\right]} \tag{2.47}$$

However, the dependence of η_{iT} on the velocity ratio $U_{\mathrm{T}} / C_{\mathrm{S}}$ is much stronger than the dependence on the pressure ratio p_a / p_E. Therefore, the turbine isentropic efficiency is approximated for modelling purposes by a second-order polynomial of the form:

$$\eta_{iT} = \kappa_{\mathrm{T02}} \cdot \left(\frac{U_{\mathrm{T}}}{C_{\mathrm{S}}}\right)^2 + \kappa_{\mathrm{T01}} \cdot \left(\frac{U_{\mathrm{T}}}{C_{\mathrm{S}}}\right) + \kappa_{\mathrm{T00}} \tag{2.48}$$

The interpolation polynomial coefficients κ_{T02}, κ_{T01} and κ_{T00} are calculated using the turbocharger manufacturer's charts for η_{iT}. Note that the η_{iT} curve for the average pressure ratio value has to be selected and then interpolated by a polynomial of the above form.

Finally, the parameter $\mathrm{a_T}$ can be approximated by a second-order polynomial of the turbine pressure ratio p_a / p_E, *i.e.*:

$$\mathrm{a_T} = \kappa_{\mathrm{T12}} \cdot \left(\frac{p_a}{p_E}\right)^2 + \kappa_{\mathrm{T11}} \cdot \left(\frac{p_a}{p_E}\right) + \kappa_{\mathrm{T10}} \tag{2.49}$$

Coefficients κ_{T12}, κ_{T11} and κ_{T10} of the interpolation polynomial for $\mathrm{a_T}$ are calculated using the specification sheets provided by the turbocharger manufacturer.

2.5 Cycle-mean Model Summary and Solution Procedure

2.5.1 Direct-drive Turbocharged Engine Model Summary

For the sake of compactness the algebraic and differential equations used for the thermodynamic engine model considered in this text are presented summarised in Table 2.2. Note that the presentation order here corresponds to the solution procedure followed and not to engine partitioning, as in the previous section, where the physical insight of the model was investigated.

Table 2.2 Engine thermodynamic model summary

I. ALGEBRAIC EQUATIONS

group A.0	
Fuel flow (eq. 2.A.0.1)	$\dot{m}_F = z_c \cdot \mathrm{m}_{\mathrm{F,max}} \cdot F_R \cdot N_E$
Scavenging pressure (eq. 2.A.0.2)	$p_I = p_a \cdot \left[\kappa_{\mathrm{C}} \cdot \left(\frac{N_{TC}}{\sqrt{T_a}} \right)^2 + 1 \right]$
group A.1	
Turbine flow correction parameter (eq. 2.A.1.1)	$\mathrm{a_T} = \kappa_{\mathrm{T12}} \cdot \left(\frac{p_a}{p_E} \right)^2 + \kappa_{\mathrm{T11}} \cdot \left(\frac{p_a}{p_E} \right) + \kappa_{\mathrm{T10}}$
Combustion efficiency (eq. 2.A.1.2)	$\eta_c = \begin{cases} 1 & \text{, if } (\mathrm{A/F}) \geq (\mathrm{A/F}) \\ \frac{(\mathrm{A/F}) - (\mathrm{A/F})_{\mathrm{low}}}{(\mathrm{A/F})_{\mathrm{high}} - (\mathrm{A/F})_{\mathrm{low}}} & \text{, if } (\mathrm{A/F})_{\mathrm{low}} < (\mathrm{A/F}) < (\mathrm{A/F})_{\mathrm{high}} \\ 0 & \text{, if } (\mathrm{A/F}) \leq (\mathrm{A/F})_{\mathrm{low}} \end{cases}$
BMEP (eq. 2.A.1.3)	$p_e = (1 - \kappa_{\mathrm{f2}}) \cdot \mathrm{p_{i,max}} \cdot \eta_c \cdot F_R + \kappa_{\mathrm{f1}} \cdot N_E + \kappa_{\mathrm{f0}}$
Scavenging temperature (eq. 2.A.1.4)	$T_I = \kappa_{\mathrm{IC}} \cdot \dot{m}_A \cdot T_a \cdot \left[\frac{\left({p_I}/{p_a} \right)^{\frac{\gamma_A - 1}{\gamma_A}} - 1}{\eta_{iC}} + 1 \right] + (1 - \kappa_{\mathrm{IC}} \cdot \dot{m}_A) \cdot T_w$

Exhaust temperature (eq. 2.A.1.5)

$$T_E = \frac{C_{\mathrm{P,air}} \cdot T_I + \dfrac{\mathrm{H_U} \cdot (\kappa_{\mathrm{Z1}} \cdot p_e + \kappa_{\mathrm{Z0}})}{(\mathrm{A/F})}}{C_{\mathrm{P,exh}} \cdot \left(1 + \dfrac{1}{(\mathrm{A/F})}\right)}$$

---

group A.it

Air flow (eq. 2.A.it.1)

$$\dot{m}_A = C_{\mathrm{V}} \cdot \bar{A}_{\mathrm{V}eq} \cdot \frac{p_I}{\sqrt{R_{\mathrm{air}} \cdot T_I}} \cdot \left\{ 2 \cdot \left(\frac{\gamma_A}{\gamma_A - 1}\right) \cdot \left[\left(\frac{p_E}{p_I}\right)^{\frac{2}{\gamma_A}} - \left(\frac{p_E}{p_I}\right)^{\frac{\gamma_A + 1}{\gamma_A}} \right] \right\}^{\frac{1}{2}}$$

Exhaust pressure (eq. 2.A.it.2a)

if $\dfrac{p_E}{p_a} < 1.85$

$$p_E = \frac{(\dot{m}_A + \dot{m}_F) \cdot \sqrt{R_{\mathrm{exh}} \cdot T_E}}{\mathrm{a_T} \cdot A_{\mathrm{T}eq}} \cdot \left\{ 2 \cdot \left(\frac{\gamma_E}{\gamma_E - 1}\right) \cdot \left[\left(\frac{p_a}{p_E}\right)^{\frac{2}{\gamma_E}} - \left(\frac{p_a}{p_E}\right)^{\frac{\gamma_E + 1}{\gamma_E}} \right] \right\}^{-\frac{1}{2}}$$

or (eq. 2.A.it.2b)

if $\dfrac{p_E}{p_a} \geq 1.85$

$$p_E = \frac{(\dot{m}_A + \dot{m}_F) \cdot \sqrt{R_{\mathrm{exh}} \cdot T_E}}{\mathrm{a_T} \cdot A_{\mathrm{T}eq}} \cdot \left\{ \gamma_E \cdot \left(\frac{2}{\gamma_E + 1}\right)^{\frac{\gamma_E + 1}{\gamma_E - 1}} \right\}^{-\frac{1}{2}}$$

---

group A.2

Exhaust gas velocity (eq. 2.A.2.1)

$$C_{\mathrm{S}} = \sqrt{2 \cdot C_{\mathrm{P,exh}} \cdot T_E \cdot \left[1 - \left(\frac{p_a}{p_E}\right)^{\frac{\gamma_E - 1}{\gamma_E}} \right]}$$

Turbine isentropic efficiency (eq. 2.A.2.2)

$$\eta_{iT} = \kappa_{\mathrm{T02}} \cdot \left(\frac{\pi \cdot D_{turb} \cdot N_{TC}}{C_{\mathrm{S}}}\right)^2 + \kappa_{\mathrm{T01}} \cdot \left(\frac{\pi \cdot D_{turb} \cdot N_{TC}}{C_{\mathrm{S}}}\right) + \kappa_{\mathrm{T00}}$$

Turbine torque (eq. 2.A.2.3)

$$Q_T = \frac{\eta_{iT} \cdot C_{\mathrm{P,exh}} \cdot T_E \cdot (\dot{m}_A + \dot{m}_F)}{2\pi/60 \cdot N_{TC}} \cdot \left[1 - \left(\frac{p_a}{p_E}\right)^{\frac{\gamma_E - 1}{\gamma_E}} \right]$$

Compressor torque (eq. 2.A.2.4)

$$Q_C = \frac{C_{\mathrm{P,air}} \cdot T_a \cdot \dot{m}_A}{\eta_{iC} \cdot 2\pi/60 \cdot N_{TC}} \cdot \left[1 - \left(\frac{p_I}{p_a}\right)^{\frac{\gamma_A - 1}{\gamma_A}} \right]$$

Engine torque (eq. 2.A.2.5)

$$Q_E = \frac{z_c \cdot V_h}{2\pi} \cdot p_e$$

Propeller load torque (eq. 2.A.2.6)

$$Q_L = K_{Q0} \cdot N^2$$

II. DIFFERENTIAL EQUATIONS

---

group DYN

Engine/propeller shaft dynamics (eq. 2.D.1):	$\dot{N}(t) = \dot{N}_E(t) = \dfrac{Q_E - Q_L}{I}$
Turbocharger shaft dynamics (eq. 2.D.2):	$\dot{N}_{TC}(t) = \dfrac{Q_T + Q_C}{I_{TC}}$

Equation grouping will be made clear when the solution procedure is presented. However, it is obvious that the complexity of the model does not lie in the dynamical part (group DYN) but in the algebraic part (groups A.x). It is evident, also, that for the algebraic part some iterative, numerical solution method must be employed. Finally, note that the equations have been modified, in comparison to their raw physical/thermodynamic form presented in the previous section. The form presented in the Table 2.2 is the one used for the numerical implementation of the engine model. From the same perspective, some intermediate variables, such as compressed air temperature (T_C), scavenging air and exhaust gas enthalpies (h_I) and (h_E), fuel chemical energy proportion in exhaust gas (ζ_α), *etc.*, have been omitted. The expressions of these intermediate variables have been incorporated directly in the next algebraic equation in the calculation sequence.

2.5.2 Engine Simulation Procedure

Operation of the direct-drive turbocharged marine Diesel engine can be simulated using the thermodynamic, quasi-steady, cycle-mean description given above. Each simulation step proceeds in the three phases shown in the Figure 2.2following schematic, which are based on the equation grouping presented previously.

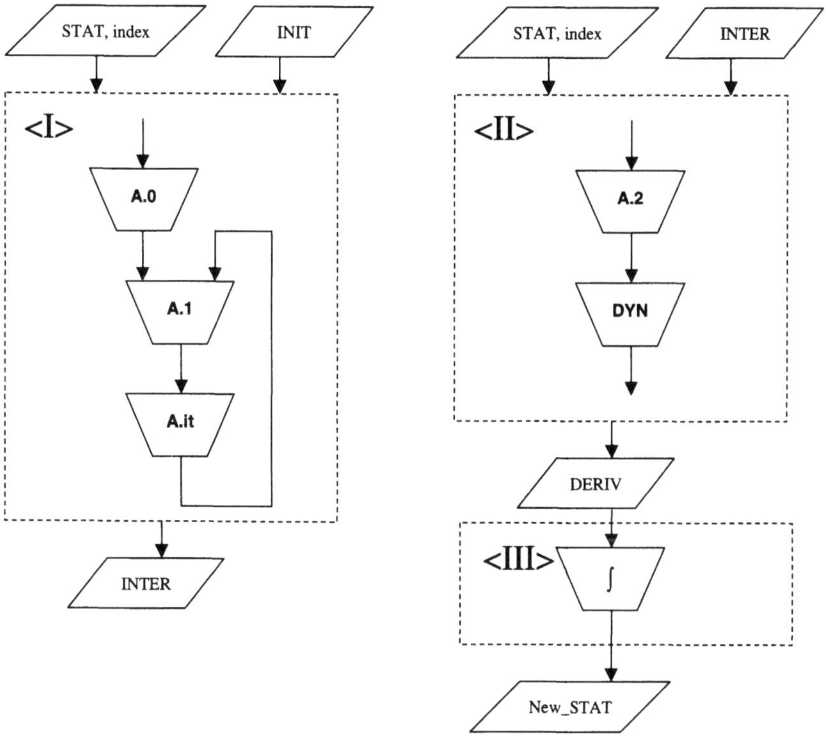

Figure 2.2 Flowchart of engine simulation procedure

In phase <I> the objective is to calculate the intermediate thermodynamic variables air mass flow rate and exhaust pressure of set INTER=$\{\dot{m}_A, p_E\}$. Specifically, starting with a set of values for the rpm variables STAT=$\{N_E, N_{TC}\}$ and the fuel index position (index $\equiv F_R$) the values of variables in set INTER can be calculated using the equations of groups A.0, A.1 and A.it.

Equations of A.0 are directly solved with respect to fuel mass flow rate $(\dot{m}_F)$ and scavenging pressure (p_I). Therefore, the values of STAT and index are propagated through A.0, providing the values for fuel mass flow rate and scavenging pressure.

On the other hand, equations of A.1 provide the values for BMEP, scavenging and exhaust temperatures, *etc.* based on the values for STAT, index and INTER=$\{\dot{m}_A, p_E\}$. Therefore, a numerical iterative procedure has to be employed for the determination of the values of variables in set INTER, as these values depend on the ones calculated from A.1 in a perplexed and non-linear system of algebraic equations. This is implemented in phase <I>. Note that an additional input set of values, namely INIT, is required for the iterative solution procedure employed for groups A.1 and A.it. This value set includes the initial guesses for

INTER=$\{\dot{m}_A, p_E\}$, the values of INIT, as well as the numerical solution method employed affect convergence to the solution. These aspects are also revisited later on.

After the values for variables of set INTER have been determined, the simulation run proceeds to phase <II>, which includes the propagation of values of STAT, index and INTER variable values through equations of group A.2, which results in the calculation of engine, turbine and compressor torque, Q_E, Q_T and Q_C respectively. In effect, the values of the torque variables allow the calculation of engine–propeller shaft and turbocharger shaft acceleration, $\dot{N} = \dot{N}_E$ and $\dot{N}_{TC}$.

Finally, in phase <III> the new values for STAT variables (new_STAT) are calculated. This is performed using a numerical integration scheme in the discrete time domain. Typical numerical integration schemes employed in control simulation practice are backward difference, forward difference and Tustin's method (trapezoidal rule). More details on numerical integration schemes are provided in [9]. Usually for engine simulations, backwards difference is employed, with time step equal to $60/(2\pi N_{MCR})$, *i.e.* the minimum duration in seconds of one engine cycle, corresponding, thus, to the time step specified for the cycle-mean model. Backwards difference has a meaningful practical interpretation, as discrete-time controllers are interfaced with Zero-Order-Hold (ZOH) D/A converter circuitry. ZOH effect is ideally represented with backwards difference [9,10]. This is why forward difference has been chosen as the integration method for the implementation of the engine model.

2.5.3 Typical Case Numerical Example

The simulation procedure described above is exemplified in the case of a marine propulsion powerplant with an MAN-B&W 6L60 marine engine. The constants required for the quasi-steady, cycle-mean thermodynamic model presented are given in Table 2.3. Note that MCR of this engine has been determined by the manufacturer to be 11,520 kW at 123 rpm [11,12]. However, the Contract MCR (C-MCR), *i.e.* the values of speed (rpm) and power according to which the engine has been matched for the ship, are significantly lower. Specifically, C-MCR is set to 9177 kW at 114.6 rpm. This is due to engine–propeller–hull matching considerations. Indeed, as quite often in marine plant design practice, the engine is overpowered in comparison to propulsion requirements, so that adequate service and sea margins are provided.

Four simulation runs were performed with "manual" fuel index step changes. Fuel index initial values of 35, 40, 45 or 50% for each one of the four simulation runs and at $t = 3$ s; this was then stepwise changed to 100% in each case. Then, the engine response in terms of shaft and turboshaft rpm, as well as the air mass flow rate and exhaust pressure (the values calculated by the perplexed non-linear system) were plotted for each case in Figures 2.3–2.6.

Table 2.3 Thermodynamic model nomenclature and typical values

p_a, T_a	Ambient (atmospheric) conditions	105, 290.0	Pa, K
R_{air}, R_{exh}	Air/exhaust gas constants	287.0, 277.1	J/kg/K
$C_{P,air}$, $C_{P,exh}$	Air/exhaust gas specific heats at constant pressure	1005.0, 1117.0	J/kg/K
γ_A, γ_E	Air/exhaust gas specific heat ratios	1.4, 1.33	-
z_c	Number of engine cylinders	6	-
$m_{F,max}$	Max fuel mass injected per cylinder and cycle	0.0393	Kg/cyl
K_{Q0}	Nominal value of propeller torque coefficient	58.838	N m/rpm^2
κ_C	Compressor map coefficient	4.0776×10^{-6}	-
η_{iC}	Compressor isentropic efficiency	0.80	-
κ_{T10}	Turbine flow correction parameter coefficients	1.6227	-
κ_{T11}		–2.2914	-
κ_{T12}		1.8006	-
$(A/F)_{low}$	Lean flammability limit	8.0	-
$(A/F)_{high}$	Perfect combustion value	17.0	-
κ_{f0}	Friction pressure coefficients	0.03	-
κ_{f1}		818.2450	-
κ_{f2}		681.3110	-
$p_{i,max}$	Max value of IMEP	16.0×10^5	Pa
κ_{IC}	Intercooler coefficient	2.0×10^{-4}	-
T_W	Coolant water temperature	30.0	°C
H_U	Fuel lower calorific value	40640	J/gr
κ_{Z0}, κ_{Z1}	Coefficients of fuel chemical energy proportion in exh. gas	1.05×10^{-7}, 0.312	-
C_V	Inlet port/exhaust valve flow discharge coefficient	0.9	-
$\bar{A}_{Veq}$	Inlet port/exhaust valve mean equivalent area	0.06	m^2
A_{Teq}	Turbine equivalent area	0.03052	m^2
D_{turb}	Turbine rotor diameter	0.5547	m
κ_{T00}	Turbine isentropic efficiency coefficients	–0.3354	-
κ_{T01}		0.9788	-
κ_{T02}		0.0875	-
V_h	Piston displacement volume	0.5429	m^3
I_0	Engine–propeller shaft nominal inertia	59800.0	kg m^2
I_{TC}	Turbocharger shaft inertia	4.83	kg m^2

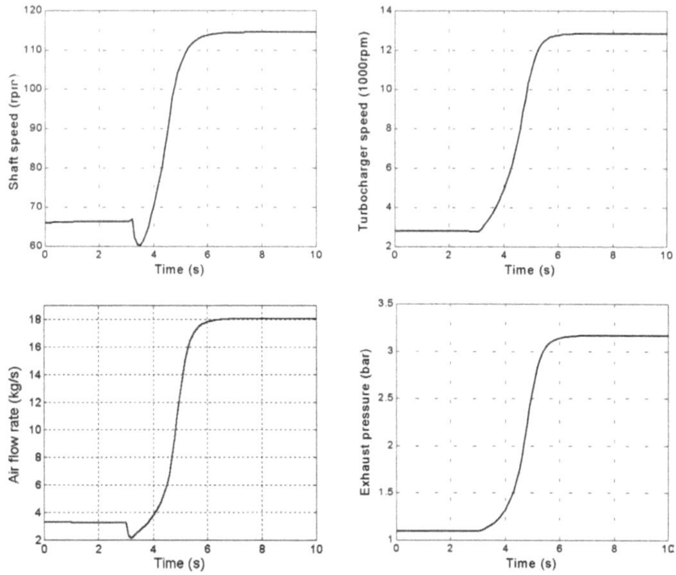

Figure 2.3 Engine transient response – 35 to 100% fuel index step at $t = 3$ s

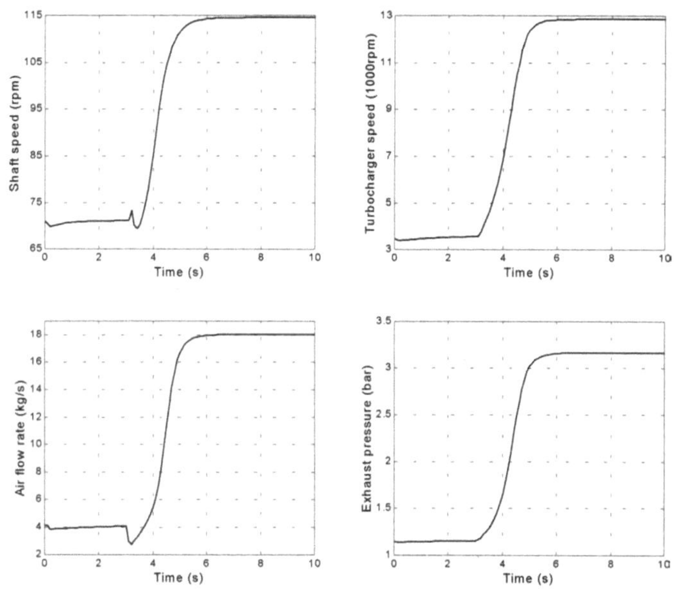

Figure 2.4 Engine transient response – 40 to 100% fuel index step at $t = 3$ s

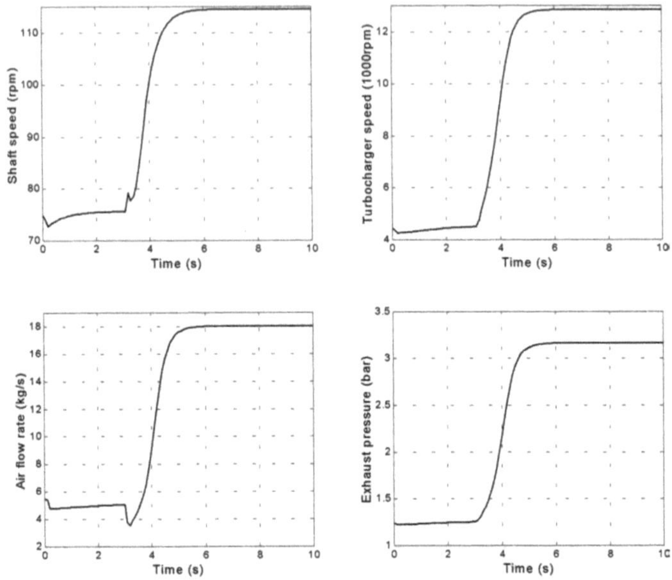

Figure 2.5 Engine transient response – 45 to 100% fuel index step at $t = 3$ s

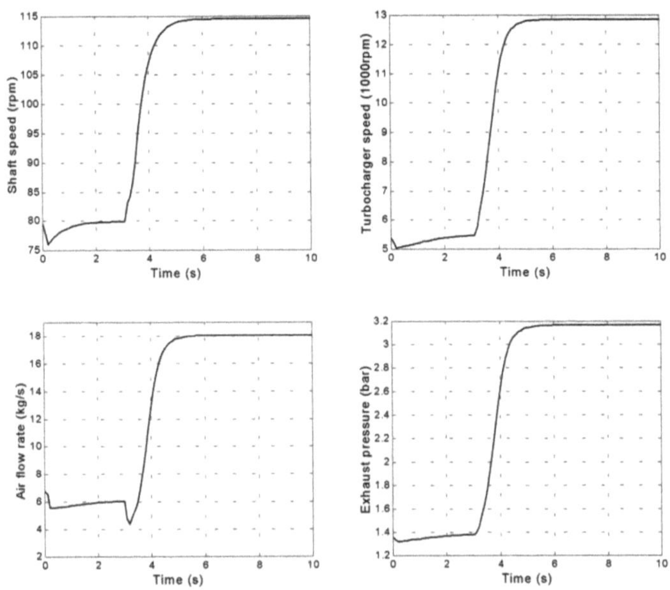

Figure 2.6 Engine transient response – 50 to 100% fuel index step at $t = 3$ s

The transient response plots depicted in Figures 2.3–2.6 demonstrate the main features of the engine behaviour. Indeed, it is readily observed that engine response, in terms of torque delivery development, is retarded to large, rapid transients, such as the ones performed in simulation. As explained in upcoming chapters this retardation also is due to the temporary inability of the turbocharging system to provide adequate combustion air mass. Eventually, however, equilibrium is restored, but only after the turbocharger(s) has accelerated to its nominal rpm corresponding to 100% engine load.

The nonlinearity of the processes involved in thermal power/torque generation is manifested in the above transient response plots. Indeed:

- the rise time for all plant signals of interest is varies greatly and is dependent upon the amplitude of the step imposed;
- the shape (pattern), especially of shaft rpm response, is dependent upon the amplitude of the response, thus implying latent dynamical terms activated under certain transient conditions.

The above remarks lead to the conclusion that, for control purposes, a non-linear model depicting the plant dynamical behaviour has to be used. In this respect, the perplexed, non-linear algebraic system of equations has to be eliminated. This is achieved with the construction and use of engine and turbocharger torque maps with the procedure presented hereafter.

2.5.4 Torque Map Generation Procedure

It should be noted here that simulation using the algebraic part of the engine model proceeds rather slowly, due to the computational effort required for the iterative solution of the algebraic system of equations consisting of groups A.1 and A.it. This difficulty can be overcome by the construction and use of torque value maps.

The separation of the engine model equations into algebraic and dynamical ones allows the solution of all algebraic equation groups (A.0, A.1, A.it and A.2) for virtually any *ad hoc* value of the triad (N, N_{TC}, F_R). Specifically, the equations of group A.it have the following generic, conceptual form, if appropriately rearranged:

$$\left.\begin{aligned} \dot{m}_A &= F_{\text{AIR}}(\dot{m}_A, p_E)\big\|_{(N,N_{TC},F_R)} \\ p_E &= F_{\text{EXH}}(\dot{m}_A, p_E)\big\|_{(N,N_{TC},F_R)} \end{aligned}\right\} \Leftrightarrow \vec{\theta}(\dot{m}_A, p_E) \triangleq \begin{bmatrix} \dot{m}_A - F_{\text{AIR}}(\dot{m}_A, p_E)\big\|_{(N,N_{TC},F_R)} \\ p_E - F_{\text{EXH}}(\dot{m}_A, p_E)\big\|_{(N,N_{TC},F_R)} \end{bmatrix} = \vec{0} \tag{2.50}$$

Note that conceptual functions $F_{\text{AIR}}(\dot{m}_A, p_E)$ and $F_{\text{EXH}}(\dot{m}_A, p_E)$ are parameterised by the value of the triad (N, N_{TC}, F_R). Their actual mathematical expression is rather long, as one has to pass through equation groups A.0, A.1 and A.it in order to obtain the explicit closed form. Therefore, it is omitted for simplicity. In the literature [13–15], a variety of numerical iterative solution methods of non-linear, perplexed algebraic systems, such as the above, are investigated. A brief resume of such methods is given in Appendix A.

In effect, torque maps for the engine, turbine and compressor torque, of the form $Q_E(N, N_{TC}, F_R)$, $Q_T(N, N_{TC}, F_R)$ and $Q_C(N, N_{TC}, F_R)$ respectively, can be constructed. The grid of points in $\mathbb{R}^3$ at which the torque functions of the above form are evaluated can be as dense as required for the simulation purposes, *e.g.* to cover the operating region of interest with adequate accuracy. Evidently, during simulation, interpolation has to be used in order to calculate torque values not explicitly included in the maps. Except for linear or quadratic interpolation schemes, neural nets can be employed for this objective. This approach provides a number of additional advantages for control system development, and this is why it is examined in depth in Chapter 5.

The standard trust region method [14] was used for the solution of the algebraic system of the engine quasi-steady, cycle-mean thermodynamic model with satisfactory results, both in terms of solution determination accuracy and in terms of speed of convergence and computational effort requirements. The achievements are demonstrated in the following typical test case for the MAN-B&W 6L60 marine engine presented previously. The torque maps produced in this manner are valuable not only for accelerating engine simulation in the computer, but also for developing plant models suitable for control purposes, as shown in Chapter 5.

2.5.5 Test Case Investigation

The MAN-B&W 6L60 marine propulsion engine paradigm is extended to demonstrate the torque map generation procedure described above. Five maps have been constructed for the following thermodynamic and mechanical variables of interest: air mass flow rate, exhaust pressure, engine torque delivery, turbocharger turbine torque delivery and compressor load torque.

The algebraic system for this specific engine has been formulated using the parameters provided in Table 2.1. A large amount of data points were mapped, covering both normal operating conditions and more or less unrealistic conditions. The unrealistic cases were considered in order to check the numerical robustness of the trust region method [14] used to solve the perplexed algebraic system. This method of large-scale optimisation has been proven to perform well in the specific problem and test case examined.

A small portion of the data set generated is shown in Figures 2.7 and 2.8. Note that all maps are $\mathbb{R}^3$-valued functions, meaning that they cannot be visualised on line or surface plots. Therefore, the presentation strategy followed was to plot the dependence of each thermodynamic or mechanical variable of interest on turboshaft rpm over a wide range of values. The shaft rpm and engine load (fuel index) are given as parameters in each case.

The major conclusion drawn from studying the plots in Figures 2.7 and 2.8 is that engine speed does not greatly affect the variables of interest. This is mainly due to the fact that the engine is two-stroke. On the other hand, turbocharger speed and engine load, in terms of fuel index, affect all variables. In the case of engine torque, the main factor is fuel index, as long as adequate combustion air is available. This, in turn, is defined by the turbocharger speed value. As can be seen, there is a "breakdown" point where engine torque ceases to be insensitive to turbocharger speed. This point corresponds to air-to-fuel ratio values lower than $(A/F)_{high} = 17.0$. Therefore, BMEP, and in effect engine torque, is becoming a linear function of A/F, leading to the maps obtained.

In the case of turbine and compressor torques, note that the two lines intersect at a point where, evidently, turboshaft equilibrium is observed. This value of turbo rpm commonly corresponds to the "flat" part of the engine torque plot. This means that the engine–propeller shaft speed is eventually stabilised at a value where the propeller absorbs the engine torque delivery, determined by the fuel index. Otherwise, the turbocharging system is stable. Indeed, above the turboshaft rpm value of equilibrium, the compressor decelerating torque is larger than the delivery of the turbine, causing the turbocharger to recover the equilibrium speed; the opposite happens for turboshaft speed values lower than the equilibrium one.

Finally, the values of air flow rate and exhaust pressure demonstrate significant dependence on turbo rpm and fuel index, as expected. They are both increasing functions of turbo rpm and fuel index. However, the rapid degradation of both these variables from their nominal value as turbo rpm is perturbed from the steady-state value is remarkable and points out the non-linear character of marine engine operation, as well as dependence of normal operating conditions on the turbocharger state.

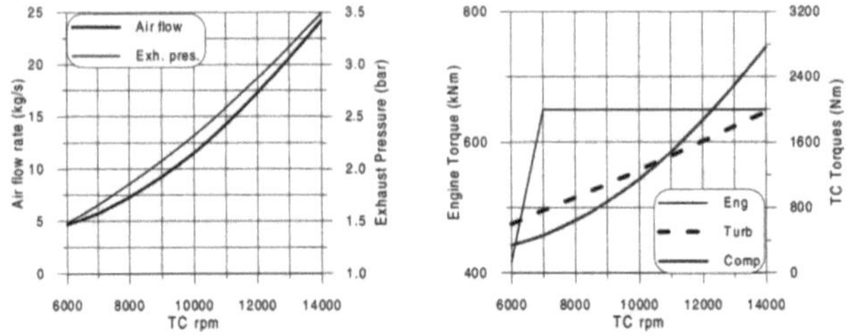

RPM = 100, %index = 85

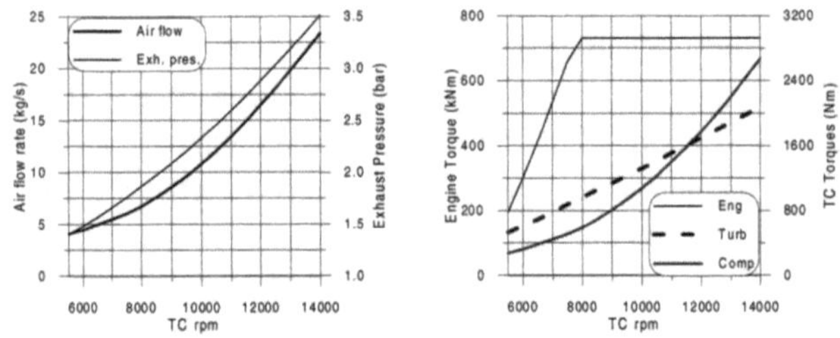

RPM = 100, %index = 95

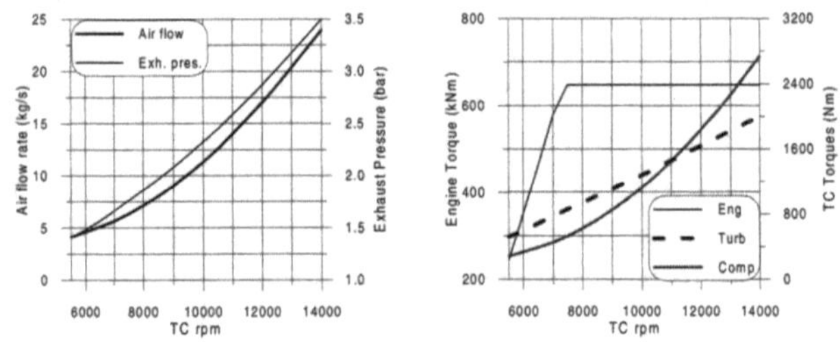

RPM = 105, %index = 85

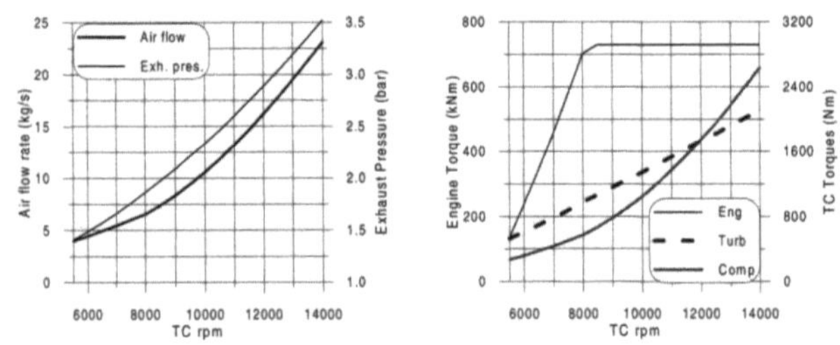

RPM = 105, %index = 95

Figure 2.7 Engine and turbocharger torque functions parameterised by rpm and index

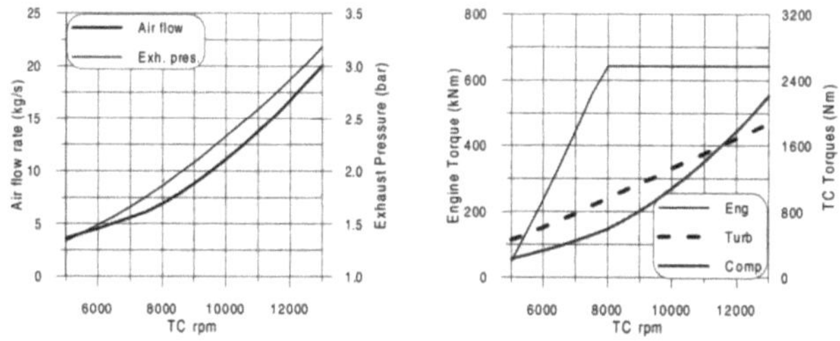

RPM = 110, %index = 85

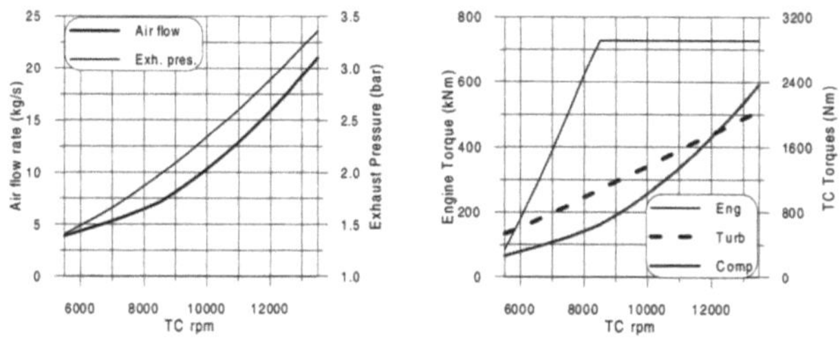

RPM = 110, %index = 95

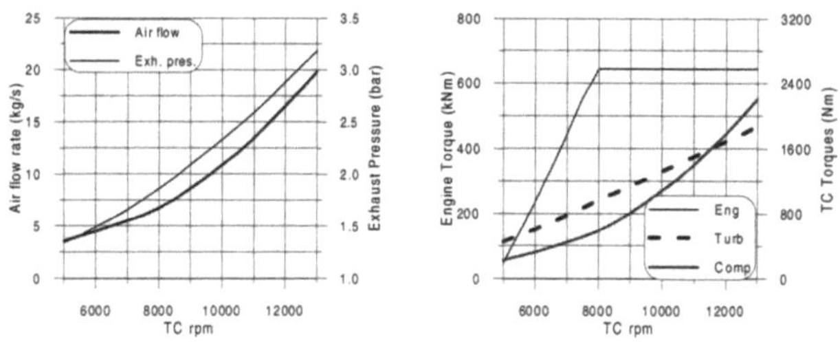

RPM = 115, %index = 85

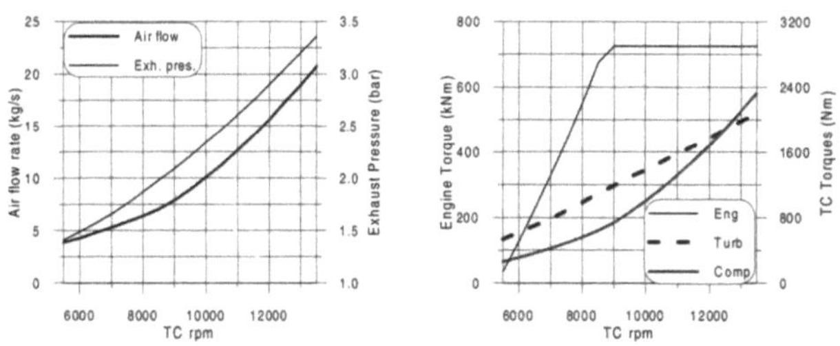

RPM = 115, %index = 95

Figure 2.8 Engine & turbocharger torque functions parameterised by rpm and index

2.6 Summary

The governing physical principles and equations of the engine thermodynamic processes of power/torque generation are presented; furthermore, the dynamical equations, derived from the mechanics of the shafts in the plant, are given. The analysis has been based on a standard quasi-steady cycle-mean thermodynamic model, which is formulated as an interconnection of the various powerplant subsystems, *e.g.* turbocharger, cylinders, intercooler, *etc.* In order to gain insight into actual engine processes, a solution and simulation procedure has been deployed based on the decomposition of the physical powerplant description to a dynamical and an algebraic (static) part. The solution procedure for the algebraic part, consisting of a non-linear and perplexed set of equations, requires the employment of numerical, iterative techniques. To bypass this, and in order to facilitate the simulation procedure, the construction of torque maps has been proposed. The construction of these maps was based on the solution in advance of the algebraic system with a specific numerical method for various values of the independent parameters involved, without taking into account the dynamical equations of shaft rotational motion. On the other hand, the dynamical part, consisting of just two first-order differential equations, is readily formulated under the modelling assumption of lumped shaft elements. In effect, the transfer function (Chapter 3) and state-space description (Chapter 5) of the marine plant will be formulated, taking into account that the order of the plant is two and that linearisation is valid around the vicinity of the steady-state operating curve.

CHAPTER 3

MARINE PLANT EMPIRICAL TRANSFER FUNCTION

3.1 Black-box Engine Modelling

In the previous chapter a physical model describing the operation of the marine engine has been presented. Based on the results obtained using such physical descriptions of the open-loop (uncontrolled) plant the dynamical behaviour and response of the plant can be studied and analysed in order to deduce requirements for feedback control and specifications for closed-loop transient performance. The outcome of such a dynamical analysis may be quantified in the form of either a state-space model or a (preferably low-order) transfer function. This state-space or transfer function model may substitute the complicated physical model description of the system under certain assumptions and usually for a small portion of the complete admissible operating region. Eventually, it can be used for controller synthesis.

In this general framework, a transfer function is established as described in the current chapter for the marine plant and a state-space model in Chapter 5. The transfer function is used for the PI(D) speed governor tuning described in the Chapter 4. Although a variety of transfer function models for the marine plant can be found in the literature, two major aspects of the marine plant operation are often neglected [16]. The first one is concerned with the turbocharger–engine interaction [17]. Specifically, as will be seen later, the turbocharger is increasing the order of plant dynamics by one, as it introduces a fast pole to the system. In practice, the effect of turbocharging on transient performance is quite significant, especially during engine acceleration. This is why, quite commonly, the fuel index is limited below a certain value, which is a function of boost (scavenging) pressure. In effect, the PI control law is allowed to operate as long as the upper limit, calculated for each value of boost pressure, is not violated. This rather crude technique of employing the boost pressure feedback signal guarantees that adequate combustion air mass is delivered to the engine cylinders. As can be seen from the equations of the two-stroke engine model, air mass flow rate depends upon the pressure ratio between exhaust and scavenging receivers, as well as scavenging pressure value. Therefore, at each steady-state operating point, by calculating the scavenging pressure value, the maximum amount of fuel can be determined so that perfect combustion is guaranteed. This amount determines, in turn, the upper limit for fuel index position that can be supported by the specific value of scavenging pressure, so that A/F is maintained above the perfect combustion threshold.

The second aspect, usually neglected when tuning state-of-the-art speed governors, is operation under extreme propeller load torque fluctuations. Specifically, due to the inherent stability of the marine propulsion system, it has often been argued that no rpm regulation is required for marine propulsion engines coupled to screw propellers (fixed rack approach to engine control) [18,19]. Stability of the marine plant originates from the rpm-dependence of the loading imposed to the main engine by the screw propeller. Furthermore, state-of-the-art governors are usually tuned on the basis of transfer functions obtained for the nominal operating conditions of the marine propulsion powerplant. To be specific, engine loading is assumed to follow the nominal propeller law; this is valid, though, only under calm sea conditions. On the other hand, when the engine is run near MCR and significant propeller load torque fluctuations occur, the control problem of operating point regulation takes its strictest form. Indeed, as MCR is located quite close to the upper bound of the allowed operating range for modern marine engines, a significant propeller torque fluctuation may drive the operating point into the barred area. In marine practice, and in order to avoid possible engine overloading, engine, and consequently ship, speed loss is employed, *i.e.* the engine operating point is deliberately lowered so that adequate operational margins are provided. In effect this is one major reason for main engine oversizing; although the propulsion power requirements can be accommodated with a smaller engine, margins have to be superimposed, so that safe operation with adequate power output is available whenever large propeller torque fluctuations occur.

In that respect, the transfer function deduced in the current chapter concerns transient response and behaviour of the marine plant around a steady-state operating point (in the typical case MCR) under operating conditions that cause severe propeller torque demand fluctuation. Therefore, an appropriate disturbance signal is introduced, incorporating the variation of propeller torque coefficient from its nominal value. Furthermore, the influence of turbocharging on transient response is considered. The transfer function will be obtained using a combination of manufacturer steady-state performance data, as well as data series predicted by physical thermodynamic models for engine operation under dynamic loading. Use of the latter is intended to substitute measurement campaigns onboard ship for obtaining transient loading data in order to calibrate the open-loop transfer function. Indeed, the thermodynamic analysis of the previous chapter indicated that a second-order transfer function should be used, instead of the first-order one that is usually employed in literature. In this chapter, a detailed filling-and-emptying thermodynamic model is employed, which can be expected to provide increased accuracy compared to the quasi-steady cycle-mean model of the previous chapter, in order to generate the time series required for transfer function calibration. Finally, note that although the simpler model of Chapter 2 could have been used, the more detailed filling-and-emptying one, presented briefly later on, is preferred because it obtains adjustable in-cycle temporal resolution and allows for more reliable simulation of the various components of the marine powerlant.

3.2 Shafting System Dynamical Analysis

3.2.1 Lumped Two-mass Model

Before proceeding to the marine plant transfer function development and identification, however, an analysis of the engine–propeller shaft dynamics is given, in order to justify the lumped-parameter approach and related approximations made in the formulation of the propulsion plant models for control purposes. The analysis is directly deployed on a transfer function basis, as the shafting system is mainly a mechanical one; therefore, the linearity assumption is rather valid. A variety of numerical simulation models exists for the analysis of the marine propulsion shafting system, including Finite Elements Methods (FEM), that depict the distributed character of shafting system dynamics by decomposing the structure into a grid of elementary massive elements interconnected by springs and/or dampers [20]. The system of coupled differential (or difference) equations generated, although linear, possesses such high orders (10,000 or more) that analytical solution is either impractical or infeasible.

However, as in the case of analogous electric circuits and networks, it is widely accepted that lumped-parameters, low-order modelling can provide adequate insight to the operation of a system under the assumption that it is linear or close to linear [20,21]. Of course, lumped-parameters models of physical systems cannot be used for conceptual or design modifications; however, if these systems are included in a larger-scale entity, then the lumped-parameters, low-order model can be used for controller synthesis.

This happens to be the case of the propulsion shafting system. Therefore, for control purposes a lumped model, consisted of two inertias and a spring will be considered [20,21]. The inertial elements represent the total engine and propeller inertia, I_E and I_P respectively. Engine inertia I_E (left side in Figure 3.1) includes the inertia of all rotating (crankshaft and cranks, flywheel, *etc.*) or reciprocating (pistons, rods, *etc.*) parts of the engine reduced to an appropriately selected point of reference and averaged over one crankshaft revolution (cycle-mean-value approach). Engine inertia is assumed not to vary from cycle-to-cycle and, therefore, is treated as a constant parameter of the propulsion system.

Propeller inertia I_P (right side in Figure 3.1) includes the inertia I_{prop} of the blades and hub averaged over one complete rotation. Additionally, I_P includes the inertia of the entrained water mass I_{water}. In naval architecture literature it can be found that I_{water} can be assigned a value of $15\% \cdot I_{prop}$. Therefore, the nominal propeller inertia I_{P0} and, in consequence, nominal total shafting inertia I_0 are:

$$I_{P0} = I_{prop} + I_{water} \approx 1.15 \cdot I_{prop} \Rightarrow I_0 = I_E + I_{P0} \tag{3.1}$$

However, this is only a nominal value because the entrained water mass varies significantly during propeller operation, especially under rough sea conditions. This effect is modelled as an additive uncertainty amount ΔI superimposed on nominal propeller inertia I_{P0} and, in consequence, shafting inertia I_0. In effect:

$$I_P = I_{P0} + \Delta I \Rightarrow I = I_E + I_P = I_E + I_{P0} + \Delta I = I_0 + \Delta I \tag{3.2}$$

Finally, the engine–propeller shaft is considered to be a fully elastic and weightless dynamic element. The zero-damping assumption is valid as a first approximation, because the power and torque values in marine propulsion are quite large and energy losses on the shafting bearings can generally be neglected. The shaft's polar moment of inertia is small compared to those of the engine and the propeller, and therefore the shaft can be considered to be weightless. However, if desired, the shaft inertia can be reduced to one or both of its ends, augmenting engine and/or propeller inertia. The stiffness of the engine–propeller shaft is determined by Hooke's law constant K_{shaft}. This parameter is calculated by the shaft's physical and geometrical configuration data as follows:

$$K_{shaft} = (\text{Modulus of elasticity}) \times \frac{(\text{Polar moment of inertia})}{(\text{Length})} \tag{3.3}$$

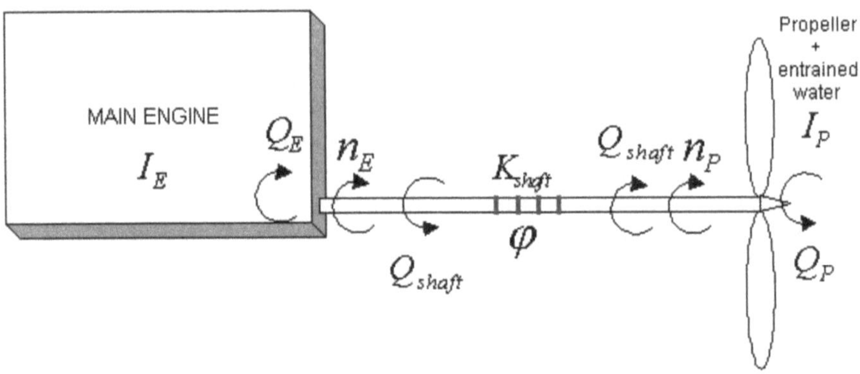

Figure 3.1 Typical configuration of marine propulsion shafting system

In the Figure 3.1, φ is the shaft's end-to-end angular deformation (with stationary reference) and Q_{shaft} is the shaft torque developed at both ends due to this deformation.

Angular deformation φ is calculated as the difference between the total arc drawn at the engine and propeller sides, *i.e.*:

$$\varphi(t) = \int_0^t (n_E(\xi) - n_P(\xi)) \cdot d\xi \tag{3.4}$$

For the above to hold, zero (temporal) initial conditions have to be assumed, *i.e.* the shaft is stationary with no loading applied to either one of its ends at $t = 0$.

By assuming zero initial conditions (initially stationary shaft) one obtains that:

$$N_{E \text{ or } P}(t) \equiv n_{E \text{ or } P}(t) \tag{3.5}$$

where:

$$N_{E \text{ or } P}(t) = N_{E \text{ or } P}(t=0) + n_{E \text{ or } P}(t) \text{ and } n_{E \text{ or } P}(t=0) = 0 \tag{3.6}$$

Furthermore, it holds that:

$$\dot{N}_{E \text{ or } P}(t) \equiv \dot{n}_{E \text{ or } P}(t) \tag{3.7}$$

By applying Hooke's law to the fully elastic shaft we obtain:

$$Q_{shaft} = K_{shaft} \cdot \varphi \tag{3.8}$$

Under the considerations presented above, the dynamical equations of motion, for the system of the elastic shaft with lumped inertias at its two ends, are as follows [20,22]:

$$\begin{aligned} I_E \cdot \dot{N}_E &= I_E \cdot \dot{n}_E = Q_E - Q_{shaft} \\ I_E \cdot \dot{N}_P &= I_P \cdot \dot{n}_P = Q_{shaft} - Q_P \end{aligned} \tag{3.9}$$

In order to obtain the canonical equations of motion [20,22] for the system the above equations are subtracted and added, yielding:

$$\begin{aligned} I_E \cdot \dot{n}_E + I_P \cdot \dot{n}_P &= Q_E - Q_P \\ I_E \cdot \dot{n}_E - I_P \cdot \dot{n}_P &= Q_E + Q_P - 2 \cdot Q_{shaft} \end{aligned} \tag{3.10}$$

The first one of the above can be easily converted to Newton's law of motion for the centre of gravity (centre of mass) of the shafting system:

$$I_E \cdot \dot{n}_E + I_P \cdot \dot{n}_P = Q_E - Q_P \Rightarrow \frac{I_E \cdot \dot{n}_E + I_P \cdot \dot{n}_P}{I_E + I_P} = \frac{Q_E - Q_P}{I} \tag{3.11}$$

The expression on the left side of Equation 3.11 can be easily identified as the rotational speed of the shaft's centre of mass and is the signal sensed by the tachometer's encoder in order to be fed to the governor, *i.e.*:

$$N \equiv n \triangleq \frac{I_E \cdot n_E + I_P \cdot n_P}{I_E + I_P} \Rightarrow \dot{n} = \frac{I_E \cdot \dot{n}_E + I_P \cdot \dot{n}_P}{I_E + I_P} \Rightarrow \dot{n} = \frac{Q_E - Q_P}{I} \tag{3.12}$$

In order to solve the system of the two coupled differential equations of motion, the Laplace transform can be employed. After algebraic manipulation and taking into account that zero initial conditions have been assumed, the Laplace transforms are as follows:

$$\begin{Bmatrix} I_E \cdot s n_E(s) = Q_E(s) - K_{shaft} \cdot \dfrac{n_E(s) - n_P(s)}{s} \\ I_P \cdot s n_P(s) = K_{shaft} \cdot \dfrac{n_E(s) - n_P(s)}{s} - Q_P(s) \end{Bmatrix}$$

$$\Updownarrow$$

$$\begin{Bmatrix} (I_E s^2 + K_{shaft}) \cdot n_E(s) - K_{shaft} \cdot n_P(s) = s \cdot Q_E(s) \\ K_{shaft} \cdot n_E(s) - (I_P s^2 + K_{shaft}) \cdot n_P(s) = s \cdot Q_P(s) \end{Bmatrix} \tag{3.13}$$

By solving the above algebraic system with respect to $n_E(s)$ and $n_P(s)$ we obtain that:

$$
\begin{aligned}
n_E(s) &= \frac{I_P s^2 + K_{shaft}}{s \cdot (I_E I_P \cdot s^2 + I \cdot K_{shaft})} \cdot Q_E(s) - \frac{K_{shaft}}{s \cdot (I_E I_P \cdot s^2 + I \cdot K_{shaft})} \cdot Q_P(s) \\
n_P(s) &= \frac{K_{shaft}}{s \cdot (I_E I_P \cdot s^2 + I \cdot K_{shaft})} \cdot Q_E(s) - \frac{I_E s^2 + K_{shaft}}{s \cdot (I_E I_P \cdot s^2 + I \cdot K_{shaft})} \cdot Q_P(s)
\end{aligned} \tag{3.14}
$$

The following remarks can be made concerning the above expressions:

- At steady-state $(s = 0)$ or close to steady-state $(s = j\omega \to 0)$ the two expressions coincide with Newton's law of motion for the centre of mass, as expected:

$$
n(s) = n_E(s) = n_P(s) = \frac{Q_E(s) - Q_P(s)}{I \cdot s} \tag{3.15}
$$

This justifies the quasi-steady approximation to ignore shafting system dynamics. Also, note that the above holds in the case of an ideally inelastic (rigid) shaft (*i.e.* $K_{shaft} \to \infty$), which is expected, too, if the rigid shaft is regarded as a degraded case of an elastic shaft element.

- The shafting system transfer function possesses three pure imaginary poles:

$$
s_0 = j0 \text{ and } s_{1,2} = \pm j\omega_c = \pm j \cdot \sqrt{\frac{I \cdot K_{shaft}}{I_E \cdot I_P}} = \pm j \cdot \sqrt{\left(\frac{1}{I_E} + \frac{1}{I_P}\right) \cdot K_{shaft}} \tag{3.16}
$$

The fact that the poles have zero real part is due to the zero-damping assumption. If some damping were to be introduced, shafting resonance points would not exhibit infinite amplitude. Furthermore, the number of poles (three) is due to the choice to model shafting dynamics with three lumped dynamic elements (two inertial and a spring element).

- For marine plants, mechanical design considerations support the following assumption [29]:

$$
I_E \approx I_P \approx I/2 \tag{3.17}
$$

Then ω_c can be estimated by the following:

$$
\omega_c \approx 2 \cdot \sqrt{\frac{K_{shaft}}{I}} \tag{3.18}
$$

By applying either the above relation or the accurate one from the lumped-parameter analysis, ω_c can be computed. For typical values for the parameters K_{shaft} and I of the marine shafting system, it can be easily seen that fuel index oscillations cannot excite any shafting harmonics, especially in the case of slow-speed, direct-drive engines. Indeed, as shown in the example below, the engine/shaft rpm values, even at MCR, are much lower than the resonance frequencies of the engine–propeller shafting system. Note that special care is taken in order that the fuel index command signal, applied to the engine, does not cause significant in-cycle variations in the fuel mass injected in each cylinder. In other words, injected fuel mass is approximately the same for all engine cylinders in the same cycle. This aspect of marine operation, however, is re-visited when governor design is discussed.

3.2.2 Typical Case Numerical Investigation

The propulsion system of a large containership is composed of an MAN-B&W 9K90MC-VI marine engine with MCR = 94 rpm and power 41 MW at MCR. The engine–propeller shaft length is 12 m and the diameter is 80 cm. The material used is stainless steel with torsional modulus of elasticity 8.2×10^{10} Pa/rad. The total engine inertia is $I_E = 444252.9 \text{ kg m}^2$ and the propeller "air" inertia $I_{prop} = 288975.0 \text{ kg m}^2$.

The propeller inertia I_P, if the entrained water mass is included, is:

$$I_P = I_{prop} + I_{water} \approx 115\% \cdot I_{prop} = 332321.0 \text{ kg m}^2$$

The total shafting system inertia I is:

$$I = I_E + I_P = 777350.0 \text{ kg m}^2$$

The polar moment of inertia of the engine–propeller shaft is calculated by the relationship:

$$(\text{Polar moment of inertia}) = \frac{\pi}{32}\times(\text{Diameter})^4 = 0.0402 \text{ m}^4 \tag{3.19}$$

Then the torsional stiffness of the engine propeller shaft is:

$$K_{shaft} = (\text{Modulus of elasticity})\times\frac{(\text{Polar moment of inertia})}{(\text{Length})} = 2.7478\cdot10^8 \text{ N m/rad}$$

In conclusion, the natural frequency of the engine–propeller shaft is:

$$f_c = \frac{1}{2\pi}\cdot\sqrt{\left(\frac{1}{I_E}+\frac{1}{I_P}\right)\cdot K_{shaft}} = 6.05 \text{ Hz}$$

If the approximate relationship is used, which is based upon the assumption that $I_E \approx I_P$, then:

$$f_c \approx \frac{2}{2\pi}\cdot\sqrt{\frac{K_{shaft}}{I}} = 5.98 \text{ Hz}$$

Therefore, the approximation error is rather small and the validity of the assumption for typical marine shafting systems is supported.

The maximum rotational speed of the engine is the value at MCR, *i.e.*:

$$f_{MCR} = \frac{N_{MCR}}{60\ (\text{s/min})} = 1.57 \text{ rps} \approx \frac{f_c}{4}$$

Therefore, the fuel index signal cannot excite any eigenmode of the shafting system provided that the fuel mass amount injected in each engine cylinder does not demonstrate any severe in-cycle variations.

3.3 The Plant Transfer Function

3.3.1 Black-box Model Development and Identification

As mentioned, engine speed regulation can be implemented by the use of PID feedback control. Actually the standard PI(D) control law forms the core linear module of the vast majority of modern electronic Diesel engine governors. PI(D) control design, however, relies on a standard form of plant (engine) model employing transfer functions and linearisation of any nonlinearities inherent in thermodynamic, physicochemical processes such as combustion. In related literature [23–25] it can be seen that this standard form includes a steady-state gain K, a single real, and, in most cases, stable, pole $(-1/\tau)$ and a dead time term τ_D. This type of dynamic behaviour is depicted in the following first-order, linear, inhomogeneous differential equation:

$$(\tau \cdot \frac{d}{dt} + 1) \cdot y(t) = K \cdot u(t - \tau_D) \tag{3.20}$$

The open-loop transfer function of the system is provided as the Laplace transform of the above equation:

$$H(s) = \frac{y(s)}{u(s)} = K \cdot \frac{e^{-\tau_D s}}{\tau s + 1} \tag{3.21}$$

A block diagram realisation of this type of dynamics is shown in Figure 3.2:

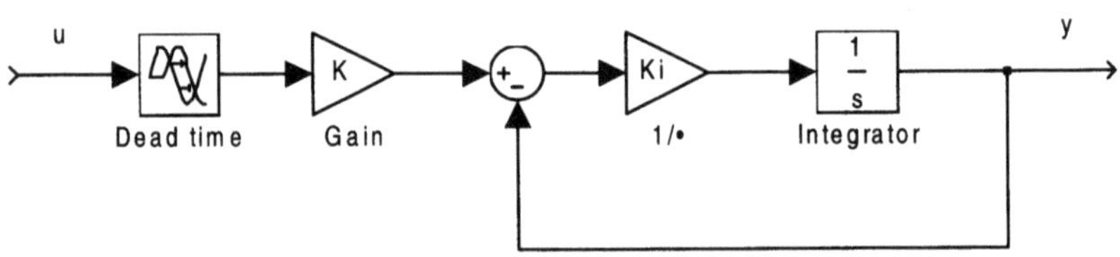

Figure 3.2 First-order system with dead time

The above form is assumed for controller design, *e.g.* when one of the Ziegler–Nichols methods (time or frequency domain) [23–25] is employed. Actually, the values for the constants of the PI(D) control law are functions of the values of the three constants involved in the above empirical dynamic model of the process, *i.e.* K, τ and τ_D. More details on PID controller tuning are provided in the literature [23–25].

However, the main problem is identification of the process dynamics. In process industry, where similar problems arise due to the complexity of the phenomena involved, a number of experiments are performed in order to obtain the required values of the open-loop transfer function. Typically, this is achieved by a number of experiments including the application of step or sinusoid excitations either as control inputs or as disturbances. Using least-squares techniques in the frequency or time domain the three constants of the transfer function can then be estimated, so that the first-order linear dynamic model response is close to that of the process. Alternatively, relay feedback approaches have been proposed and been applied extensively in the chemical industry for the identification of plant dynamics [23–25].

In the case of marine propulsion plants, and due to commercial reasons, as well as cost and insurance issues (*e.g.* strict trading schedule of the ship), the above experimental techniques cannot be readily used in many cases. Therefore, the dynamical behaviour of the marine propulsion plant can be investigated either by the use of manufacturer data, which are usually steady-state performance data, or by the use of physical models for performance prediction. In the following of the current section, an overview of the empirical transfer function models that can be used for tuning the PI speed governor is given.

3.3.2 Full-order Transfer Function

Two types of transfer function will be used for PI speed governor design and validation: the full-order transfer function and the reduced-order one. The full-order transfer function is formulated here, from the perspective presented in Section 3.1; then, the order of the plant dynamics is appropriately reduced to unity by neglecting dynamical terms. This is done in order to facilitate PID controller synthesis, as described in Chapter 4.

In the case of marine powerplants the order of the transfer function is determined by the integrating action of the engine–propeller shaft, depicted in the following dynamical equation of motion:

$$\dot{N}(t) = \frac{Q_E(t) - Q_L(t)}{I} \tag{3.22}$$

where N is the shaft rpm, Q_E is the engine torque delivery, Q_L is the propeller torque demand and I is the total shafting system inertia accounting for engine, and propeller plus entrained water inertias. In detail, inertia I can be calculated as the sum:

$$I = I_E + I_P + \Delta I \tag{3.23}$$

where I_E is the engine plus shaft inertia, I_P is the "air" propeller inertia I_{prop} augmented by a nominal 15% accounting for the entrained water inertia I_{water}. Inertia amount ΔI stands for the real parametric uncertainty introduced in I due to the variation of the entrained water mass. Note that this variation lies in the range 5–30% of propeller "air" inertia I_{prop}. Also note that the shaft dynamics of order higher than unity are neglected. This assumption has been proven in practice to hold if the cycle-mean-value variance is investigated.

Modelling of the engine thermodynamic process that determines indicated engine brake torque Q_{eng}, in transfer function plant models, consists of a thermodynamic gain C plus dead time τ_d and time constant τ_{TC} in the following first-order inhomogeneous linear differential equation:

$$\tau_{TC} \cdot \dot{Q}_{eng}(t) + Q_{eng}(t) = C \cdot F_R(t - \tau_d) \tag{3.24}$$

with initial conditions at $t = 0$:

$$Q_{eng}(0) = C \cdot F_R(0) \text{ and } F_R(t) = F_R(0), \ -\tau_d \le t \le 0 \tag{3.25}$$

where F_R is the fuel index (rack) position. Note that:

$$Q_{eng}(t) = Q_E(t) + Q_{fr} \tag{3.26}$$

where Q_{fr} is a surplus load due to mechanical losses, referred to as engine friction torque in literature, and Q_E is engine torque delivery, refered to by the term brake torque. Q_{fr} is actually dependent upon steady-state engine load expressed in either terms of fuel index or engine rpm. However, for speed regulation around an operating point it can be assumed to be approximately constant.

As shown in [26], the time constant τ_{TC} is mainly due to the effect of turbocharging on the power/torque generation process. Indeed, combustion efficiency depends mainly on the combustion (charge) airflow rate, especially at high engine loads (near MCR). This variable is controlled by the turbocharger speed, and this is why in thermodynamic cycle-mean-value models, as in the one in Chapter 2, the turbocharger rpm is included as a state variable [16]. In simple transfer function models, which are tuned using experimental data series such as the ones presented in [26], this effect can be modelled as a first-order system with a time constant determined from the time response plots. Typical transient time response plots can be found in [16,26,27].

The dead time τ_d in the engine torque equation is originating from the injection delay of the fuelling system and has been estimated in [28] to lie within the range:

$$15/N < \tau_d < 15/N + 60/(N \cdot z_c) \tag{3.27}$$

z_c number of engine cylinders.

By solving the differential equation for the indicated engine torque Q_{eng} the following expression for the engine brake torque Q_E can be obtained for $t > 0$:

$$Q_E(t) = C \cdot F_R(0) \cdot e^{-t/\tau_{TC}} + \frac{C}{\tau_{TC}} \cdot \int_0^t e^{-(t-t')/\tau_{TC}} \cdot F_R(t' - \tau_d) \cdot dt' - Q_{fr} \tag{3.28}$$

For applying the linear perturbation analysis the fuel index signal $F_R(t)$ is decomposed to a steady-state (subscript 0) plus perturbation (lower-case) signal as follows:

$$F_R(t) = F_{R0} + f_R(t) \tag{3.29}$$

By employing the above decomposition to the integral equation for the engine brake torque Q_E a similar decomposition scheme is obtained for this variable as well:

$$Q_E(t) = C \cdot F_{R0} + \frac{C}{\tau_{TC}} \cdot \int_0^t e^{-(t-t')/\tau_{TC}} \cdot f_R(t'-\tau_d) \cdot dt' - Q_{fr} \tag{3.30}$$

The initial conditions, assumed for the above integral equation, are:

$$F_R(t) = F_R(0) = F_{R0}, \ -\tau_d \le t \le 0 \text{ and } f_R(t) = f_R(0) = 0, \ -\tau_d \le t \le 0 \tag{3.31}$$

Using the Laplace transform:

$$Q_E(s) = \frac{C \cdot e^{-\tau_d s}}{\tau_{TC} s + 1} \cdot f_R(s) + \frac{C \cdot F_{R0} - Q_{fr}}{s} = \frac{C \cdot e^{-\tau_d s}}{\tau_{TC} s + 1} \cdot f_R(s) + \frac{Q_{E0}}{s} \tag{3.32}$$

where obviously:

$$Q_{E0} = C \cdot F_{R0} - Q_{fr} \tag{3.33}$$

Speed (rpm) dependent load torque is another major feature of marine propulsion plants. For marine propulsion systems a loading scheme with exceptional importance is propeller-law loading. Propeller-law loading is defined by the following equation:

$$Q_L = K_Q \cdot N^{\Xi} \tag{3.34}$$

The value of exponent Ξ in the above is assumed to be 2.0 in this text, corresponding to fully bodied hulls [29]. Torque coefficient K_Q is varying significantly and is the main source of shaft rpm fluctuation. Actually, as will be seen later on, minimisation of the effect of K_Q fluctuation on shaft rpm is the design objective of the H∞ loop shaping.

In order to deduce a transfer function of the plant transient operation, and as the propeller-law loading introduces nonlinearity, a signal decomposition scheme, similar to the fuel index case, is applied for N and K_Q:

$$N(t) = N_0 + n(t), \ K_Q(t) = K_{Q0} + k_Q(t) \tag{3.35}$$

Note that torque coefficient K_Q is treated as a signal and not as a constant parameter of the model. As explained later, perturbation $k_Q(t)$ is the major disturbance to marine plant operation and the source of perturbations for the engine rpm $N(t)$ and fuel index $F_R(t)$ signals.

Also, in order to extend the propeller law for transient operation, the following first-order Taylor approximation for Q_L is applied:

$$Q_L(t) = K_Q \cdot N^2 \approx K_{Q0} \cdot N_0^2 + 2K_{Q0} N_0 \cdot n(t) + N_0^2 \cdot k_Q(t) \tag{3.36}$$

or alternatively:

$$Q_L(t) \approx 2K_{Q0}N_0 \cdot N(t) + N_0^{\ 2} \cdot K_Q(t) - 2K_{Q0}N_0^{\ 2} \tag{3.37}$$

Using the Laplace transform we obtain that:

$$Q_L(s) = \frac{K_{Q0} \cdot N_0^{\ 2}}{s} + 2K_{Q0}N_0 \cdot n(s) + N_0^{\ 2} \cdot k_Q(s) \tag{3.38}$$

For the steady-state values (subscript 0 in the decomposition equations) propeller-law, steady-state, engine-load matching is applied so that engine torque and power delivery is equal to propulsion power requirements translated to propeller torque demand, *i.e.*:

$$Q_{E0} = Q_{L0} \Leftrightarrow C \cdot F_{R0} = K_{Q0} \cdot N_0^2 + Q_{fr} \tag{3.39}$$

where obviously:

$$Q_{L0} = K_{Q0} \cdot N_0^2 \tag{3.40}$$

In turn, this means that:

$$\dot{N}(t) \equiv \dot{n}(t) \tag{3.41}$$

Taking into account the considerations presented above, the plant dynamical equation for engine rpm perturbation $n(t)$ is as follows:

$$(I_0 + \Delta I) \cdot \dot{N}(t) =$$

$$I \cdot \dot{n}(t) = C \cdot \int_0^t \frac{e^{-(t-t')/\tau_{TC}}}{\tau_{TC}} \cdot f_R(t' - \tau_d) \cdot dt' - 2K_{Q0}N_0 \cdot n(t) - N_0^{\ 2} \cdot k_Q(t)$$

$$\Updownarrow$$

$$(I\frac{d}{dt} + 2K_{Q0}N_0) \cdot n(t) = C \cdot \int_0^t \frac{e^{-(t-t')/\tau_{TC}}}{\tau_{TC}} \cdot f_R(t' - \tau_d) \cdot dt' - N_0^{\ 2} \cdot k_Q(t) \tag{3.42}$$

Using the Laplace transform and after algebraic manipulation, it can be seen that the open-loop transfer function of the typical marine propulsion plant is as follows:

$$n(s) = \frac{C \cdot \exp(-\tau_d s)}{(\tau_{TC}s + 1) \cdot (Is + 2K_{Q0}N_0)} \cdot f_R(s) - \frac{N_0^2}{Is + 2K_{Q0}N_0} \cdot k_Q(s) \tag{3.43}$$

or in compact form:

$$n(s) = G_1(s) \cdot f_R(s) + G_2(s) \cdot k_Q(s) \tag{3.44}$$

Note that:

$$G_1(s) = \frac{n(s)}{f_R(s)} = \frac{C \cdot \exp(-\tau_d s)}{(\tau_{TC}s + 1) \cdot (Is + 2K_{Q0}N_0)} \tag{3.45}$$

has the standard form for Diesel prime movers with rpm-dependent loads, encountered in related literature such as in [26,30].

Frequently the dynamics of $G_1(s)$ are enriched so as to include the fuel index actuator. In control theory and practice, actuator dynamics are modelled as a first-order differential equation. Following this approach, the transfer function of the actuator is:

$$G_{act}(s) = \frac{1}{\tau_{act} \cdot s + 1} \tag{3.46}$$

This can be obtained from the generic transfer function $H(s)$ introduced previously for any first-order system with unity steady-state gain K, zero dead-time τ_D and time constant τ_{act}. Therefore, the transfer function $G_1(s)$ from fuel index perturbation $f_R(s)$ to shaft rpm perturbation $n(s)$ is:

$$G_1(s) = \frac{n(s)}{f_R(s)} = \frac{C \cdot \exp(-\tau_d s)}{(\tau_{act} s + 1) \cdot (\tau_{TC} s + 1) \cdot (Is + 2K_{Q0} N_0)} \tag{3.47}$$

All poles introduced in $G_1(s)$ and $G_2(s)$ are real and stable. However, dead time τ_d in $G_1(s)$ introduces a zero in the right half s-plane, if the following Taylor approximation for dead-time terms is adopted:

$$\exp(-\tau_d s) \approx 1 - \tau_d s \tag{3.48}$$

However, using this approximation, $G_1(s)$ becomes a non-minimum-phase transfer function.

The block diagram of the full-order transfer function of the propulsion powerplant is given in Figure 3.3.

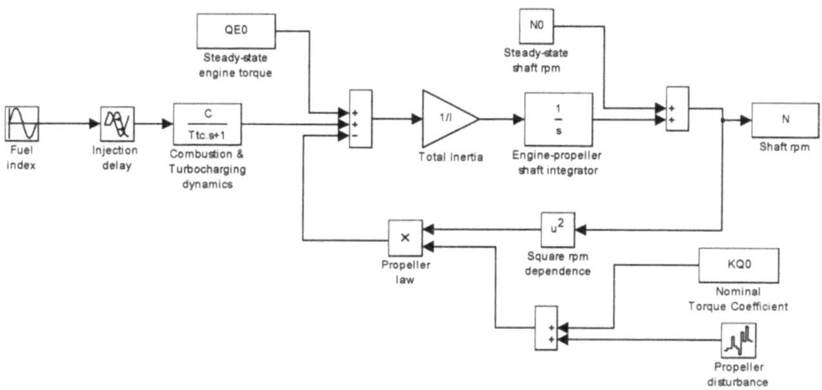

Figure 3.3 Propulsion powerplant full-order transfer function

3.3.3 Reduced-order Transfer Function

Although the order of $G_1(s)$ is three, the dominant pole is the one originating from the linearisation of the propeller law, introducing the time constant $\tau_{prop} = I/(2K_{Q0}N_0)$. This is justified by the typical numerical values of the constants involved, *i.e.* shafting system total inertia I and propeller torque coefficient K_{Q0}, as well as by the fact that ship propulsion engines are slow-speed (therefore, N_0 is comparatively small in the entire operating range, giving larger response times τ_{prop}).

In order to facilitate controller synthesis by reducing the order of the characteristic polynomial of the closed-loop system, the fast plant dynamics incorporated in $G_1(s)$ may be neglected. However, stability and performance robustness analysis of the controlled plant against the neglected plant dynamics is necessary, and this is why formulation of the full-order transfer function has been effectuated.

Fast dynamics in the case of marine plants include the following terms: (a) the turbocharging/combustion term $(1+\tau_{TC}s)$ in the full-order transfer function derived previously for the marine plant; (b) the actuator term $(1+\tau_{act}s)$; (c) the injection dead-time term $\exp(\tau_d s)$ in the augmented (full-order) plant transfer function; (d) high-order eigenmodes due to the elasticity of the engine–propeller shaft not included in the open-loop transfer function; as demonstrated by the lumped-parameters analysis, given previously, the principal (first-order) harmonic is determined by the quantity $(2\cdot\sqrt{K_{shaft}/I})$, where K_{shaft} is the torsional stiffness of the engine–propeller shaft.

Justification of the above simplifications is given now. At first, neglecting terms (a) and (b) is based upon their numerical values. Specifically, turbocharging time constant τ_{TC} can be estimated from time response plots of shaft rpm, as the ones presented in [16,26]. This represents the delay appearing mainly during rapid engine acceleration, due to the temporarily inadequate combustion air flow-rate, which is translated to retardation of the engine to develop the required torque. τ_{TC} varies greatly various engine loads and operating regimes and should be assigned a range of possible values rather than a unique value. This is due to the nonlinearities inherent in the engine physicochemical processes. Furthermore, as explained in Chapter 5, under certain conditions the pole due to turbocharging also vanishes, as it becomes unobservable (zero-pole cancellation).

The device or engine manufacturer usually provides actuator time constant and any additional time response specifications. However, advances in actuator mechatronic technology have allowed for the development of quite fast actuators (rather small τ_{act}). This supports the assumption of neglecting the associated term $(1+\tau_{act}s)$ corresponding to actuator dynamic behaviour.

Neglecting terms (c) and (d) is justified by the cycle-mean modelling approach. Indeed, shafting system vibrations, due to possible excitation of high-frequency eigenmodes, have a cycle mean value that is close to zero, especially if the operation is kept close to steady-state by means of a regulator. Here, it is to be remembered that, for two-stroke engines, such as the ones used for ship propulsion, the period of crankshaft rotation coincides with the thermodynamic cycle duration.

The same argument of cycle-mean estimation and approximation can be used in the case of the injection dead-time term. Indeed, fuel injection takes place once per cycle in each engine cylinder. Therefore, the lag introduced before fuel injection does not affect transient response, provided that the time constant of propeller load fluctuation is sufficiently greater than the period of crankshaft rotation, *i.e.* $(1/N)$. The nature and spectral content of both propeller and engine torque excitations is postponed until the next chapter, because it is closely related to the speed (rpm) regulation problem of marine plant control.

In effect, after omitting the negligible dynamical terms listed above, the reduced-order plant transfer function becomes:

$$n(s) = \frac{C}{Is + 2K_{Q0}N_0} \cdot f_R(s) - \frac{N_0^2}{Is + 2K_{Q0}N_0} \cdot k_Q(s) \qquad (3.49)$$

which gives the single, stable, real and dominant pole determined by the shaft inertia and the rpm-dependent propeller loading. The block diagram is shown in Figure 3.4.

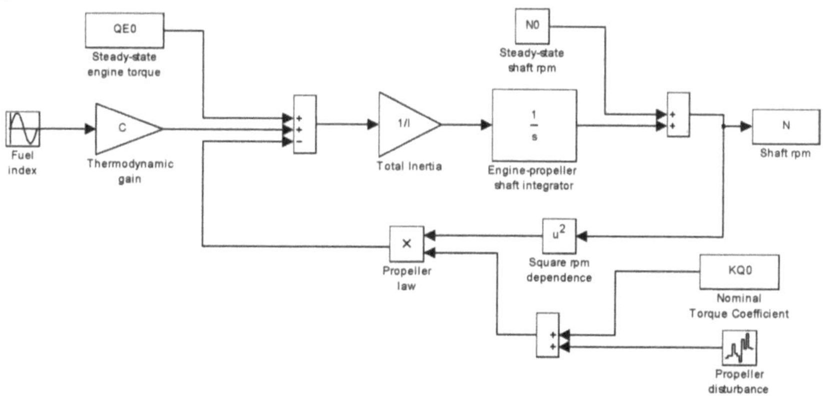

Figure 3.4 Propulsion powerplant reduced-order transfer function

This reduced transfer function demonstrates the major features of Diesel ship propulsion transient operation, and this is why it is used extensively in related literature, such as in [27]. The important conclusions are:

- The Diesel marine propulsion plant is an inherently stable system, with a pole located at $-\tau_{prop}^{-1} = (-2K_{Q0}N_0 / I)$. This is due to rpm-dependent propeller loading, as also indicated in [16].
- Despite the inherent stability, the strict requirements imposed on rpm regulation in terms of time response when the engine is run at near-MCR cannot be met without feedback fuel index control. This also is investigated in proper depth in the next chapter.

3.3.4 Plant Transfer Function Identification

As indicated in the above analysis, the transfer function models used for marine engine control include a number of parameters in the form of time constants. Manufacturer data can be used in order to estimate their numerical values. However, as explained earlier, accurate estimation of these constants cannot be done unless certain open-loop or closed-loop experiments are performed on the actual installation. For marine plants, great problems arise especially for the turbocharging/combustion time constant τ_{TC}, because, in addition to the practical difficulties, the nonlinearity of the processes involved also introduces large variations under certain conditions, and makes this pole become unobservable.

On the other hand, the dominant pole, determined by the time constant τ_{prop}, can be readily determined by the geometrical data of the propulsion shafting system. Indeed, as $\tau_{prop} = I/2K_{Q0}N_0$, the calculation of τ_{prop} requires knowledge of I, N_0 and K_{Q0}. It has been explained how the shafting inertia can be readily estimated from the engine, propeller and entrained water inertia values. The uncertainty introduced has also been identified. Furthermore, the rpm range of the engine is provided by the manufacturer, so the range of N_0 can be considered known.

Estimation of the nominal value of propeller torque coefficient K_{Q0} from the propeller geometrical data is a more cumbersome process. However, it is completely tractable, as can be seen in naval architecture literature, such as [29]. In this context, propeller torque coefficient K_{Q0} is regarded as a performance characteristic that can be estimated with required accuracy on the basis of purely geometrical data such as blade number, blade-to-area ratio, propeller pitch and diameter, *etc*. Systematic methods, based on curve-fitting of towing tank test experimental data series have been developed, that provide approximation polynomials for standard fixed-pitch propeller series. The most typical example of this methodology is the Wageningen B-screw series that can be found in [29]. Similar polynomial series are the Japanese AU, Gawn, Newton–Rader series, *etc*. However, as already explained, the actual value for propeller torque coefficient is demonstrating significant fluctuation, especially under dynamic conditions. In that respect, a main objective of propulsion feedback control is to minimise the effect of these fluctuations on normal propulsion system operation.

Therefore, the major difficulties in the identification of the marine plant transfer functions focus on assessing the range of values for the thermodynamic gain value C and the turbocharging time constant τ_{TC}, which determine the engine transient performance and dynamical behaviour. For the thermodynamic gain C a starting value can be the one obtained from the steady-state performance curves. From such typical curves of marine propulsion engines it can be easily seen that the power at steady-state operating points is a linear function of the (rated) fuel index position, which determines the engine load. *I.e.*:

$$P_{\text{brake,SS}} = \kappa_{\text{brake,1}} \cdot F_{R,\text{SS}} + \kappa_{\text{brake,0}} \tag{3.50}$$

where $F_{R,\text{SS}}$ is the corresponding steady-state index position. Parameters $\kappa_{\text{brake,1}}$ and $\kappa_{\text{brake,0}}$ can be readily determined using least-squares fitting on the manufacturer's steady-state performance curve for brake power. In turn, the engine brake torque at each steady-state operating point $Q_{E,\text{SS}}$ can be calculated as:

$$Q_{E,\text{SS}} = \frac{60}{2\pi} \cdot \frac{P_{\text{brake,SS}}}{N_{E,\text{SS}}} = \frac{60}{2\pi N_{E,\text{SS}}} \cdot (\kappa_{\text{brake,1}} \cdot F_{R,\text{SS}} + \kappa_{\text{brake,0}}) \tag{3.51}$$

Therefore, engine brake torque values are known at the steady-state operating points. Because marine propulsion engines do not operate at constant speed, the relationship between steady-state brake torque $Q_{E,\text{SS}}$ and index position $F_{R,\text{SS}}$ is not linear, due to the variation of engine rpm $N_{E,\text{SS}}$. However, direct-drive marine engines operate on the nominal square-law of the propeller, which is translated to a cubic power relationship for brake power. Therefore, relatively small changes in engine rpm result in large torque and power changes. In conclusion, as will be seen in the numerical example given below, although the brake torque–fuel index curve is not linear, it is quite close to a straight line. By using least-squares curve fitting, the equation of the straight line best-fit for the steady-state torque–index pairs is obtained. The slope of this straight line is a first approximation to the value of the thermodynamic gain *C*. However, the value obtained is rather crude; a typical error in the range 10–15% of the exact value can be expected.

A more accurate estimation of C can be obtained using thermodynamic arguments. Specifically, engine brake power P_{brake} can be calculated as the ratio between the effective work per cycle and per cylinder due to gas expansion, $W\!\!\!-_e$, and the engine cycle period, as determined by the cycle-averaged engine speed N_E [2]. In effect:

$$P_{\text{brake}} = z_c W\!\!\!-_e \cdot \frac{N_E}{60} \tag{3.52}$$

and furthermore:

$$Q_E = \frac{60}{2\pi} \cdot \frac{P_{\text{brake}}}{N_E} = \frac{z_c}{2\pi} \cdot W\!\!\!-_e \tag{3.53}$$

Effective work $W\!\!\!-_e$ can be assumed to obtain (locally in the vicinity of any steady-state operating point) a proportional relationship with fuel index position F_R, as the latter determines the fuel mass injected in the cylinders per cycle. Note that this statement holds for constant engine speed and, therefore, is not in contradiction with the statement that steady-state torque is not proportional to steady-state fuel index values, as the steady-state rpm values are different from one operating point to another. In effect:

$$\dot{W}_e \propto F_R \Rightarrow Q_E \propto F_R \tag{3.54}$$

Accurate estimation of the constants of the above proportionalities requires thermodynamic calculations. Therefore, detailed physical thermodynamic models can be used for the estimation of these proportionality constants, with required accuracy, around each operating point. In effect, thermodynamic gain C is calculated. Thermodynamic models fall into two main categories:

- Quasi-steady models, such as the one presented in the previous chapter
- Simulation codes of "control volume" (filling and emptying) type, such as Motor Thermodynamics (MoTher), which is the main tool used by the Laboratory of Marine Engineering / National Technical University of Athens for predicting the behaviour of large two-stroke marine Diesel engines and their components under steady state, as well as dynamic conditions.

A brief description of the MoTher code is given hereafter, and then an example is given in which it is demonstrated how such a simulation model can be used for estimating thermodynamic gain C, as well as turbocharging time constant τ_{TC}.

A propulsion powerplant is represented in MoTher by employing a number of basic engine elements such as flow receivers (cylinders, plenums), flow controllers (valves, compressors, turbines, heat exchangers) and mechanical elements (crankshaft, shafts, gearboxes, clutches, shaft loads). A turbocharged engine can be modelled as several flow receiver elements (control volumes) interconnected by flow controller elements. The engine's operating environment is regarded as a fixed fluid element (constant pressure, temperature and composition). In addition, mechanical connections are allowed between cylinders and crankshaft, which in addition can be connected to a shaft load via a gearbox and/or clutch.

The flow receivers are treated as open thermodynamic systems, and the work, heat and mass transfer that take place through their boundaries are calculated by applying the conservation equations in appropriate form to each control volume. The resulting system of coupled differential equations is numerically solved step-by-step for all volumes.

Sub-models for the simulation of the processes inside the control volumes, such as for combustion, heat transfer, scavenging and friction, are used. All the engine subsystems affecting the engine thermodynamics, including turbochargers compressors and turbines, charge air coolers, scavenging blowers, *etc*., are modelled. The code can predict both the variation of micro-parameters (such as the in-cylinder pressures and temperatures throughout an engine cycle), and the cumulative macro-parameters (such as heat lost, work, mean effective pressure, *etc*.) in detail for various engine configurations.

For the prediction of the engine transient response, models for the engine governor, the starting air system, and the engine/turbocharger dynamics, are also used. Setting the desirable engine speed, the output of the governor is the fuel index (rack) position, which in turn determines the mass of fuel injected into the cylinders. The instantaneous engine torque output is calculated and the engine and load torque in combination with the total inertia of the engine and load are used for calculating the rate of engine speed change.

3.3.5 Identification of Typical Powerplant

The propulsion powerplant of the 4600 TEU containership "Shanghai Express" operated by the German container liner Hapag–Lloyd is considered as a typical case of a merchant ship propulsion system. The details and steady-state performance data of the plant are summarised Figure 3.5.

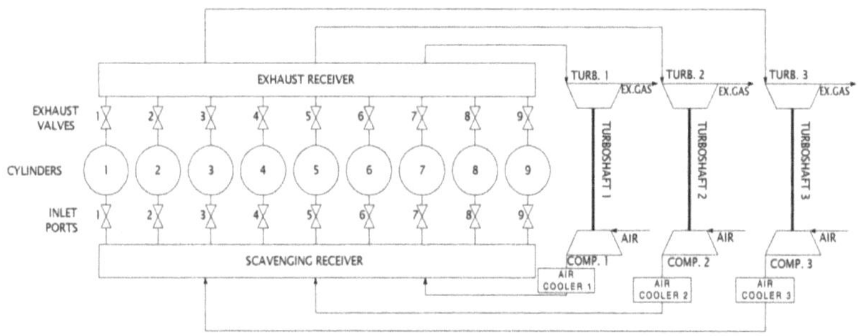

MAN B&W 9K90MC-VI - Engine Specifications

Number of cylinders	9	MCR speed	94 RPM
Bore	900 mm	MCR power	55980 BHP
Stroke	2550 mm	Boost pressure	3.6 bar
Connecting rod length	3510 mm	Brake mean effective pressure	18 bar
Compression ratio	16.762	Turbochargers	3 ABB VTR 714E-32

Figure 3.5 Hapag–Lloyd's containership "Shanghai Express" propulsion plant specifics

The propulsion plant consists of an MAN B&W 9K90MC-VI main engine with three ABB turbochargers. Except MCR, six commonly provided operating points (25, 40, 50, 75, 85 and 100% MCR) of the plant are shown Table 3.1 [31,32]. The torque values shown have been calculated using the brake power and rpm values at each operating point.

Table 3.1 Steady-state performance data of the "Shanghai Express" powerplant

Fuel index (rated) $F_{R,SS}$ (%MCR)	Engine speed $N_{E,SS}$ (rpm)	Engine brake power $P_{brake,SS}$ (MW)	Engine brake torque $Q_{E,SS}$ (10^3 kN m)
25	59.2	10.5256	1.6978
40	69.3	16.7619	2.3097
50	74.6	20.9042	2.6759
75	85.4	31.1239	3.4802
85	89.0	35.1703	3.7736
100	94.0	41.3228	4.1979

The power and torque propeller curves over which the engine is operating on steady-state are shown in Figure 3.6.

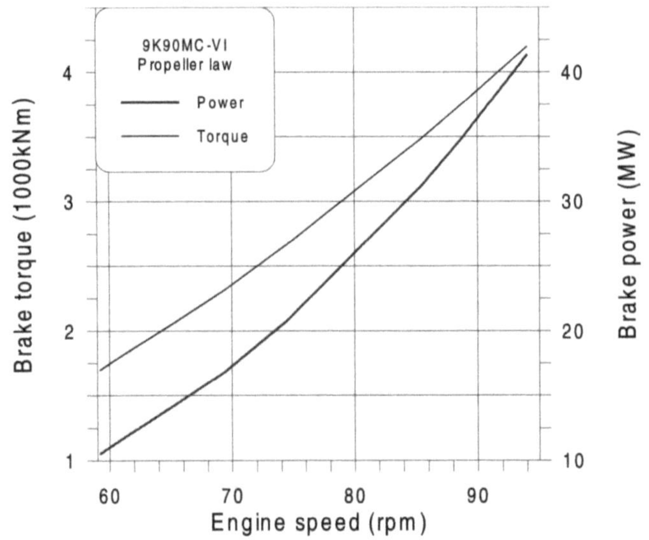

Figure 3.6 "Shanghai Express" powerplant steady-state performance curves

The use of least-squares curve-fitting gives the following polynomial relationship for steady-state brake power and torque:

$$P_{brake,SS} = K_{pow} \cdot N_{E,SS}^{\ 3} = 0.04276\ (\text{kW/rpm}^3) \cdot N_{E,SS}^{\ 3}$$

$$Q_{E,SS} = K_Q \cdot N_{E,SS}^{\ 2} = 0.4396\ (\text{kN m/rpm}^2) \cdot N_{E,SS}^{\ 2}$$

From the theoretical relationship between steady-state brake power $P_{brake,SS}$ and torque $Q_{E,SS}$ it can be readily concluded that:

$$P_{brake,SS} = Q_{E,SS} \cdot \frac{2\pi}{60} \cdot N_{E,SS} \Rightarrow K_{pow} = \frac{2\pi}{60} \cdot K_Q \approx \frac{K_Q}{10} \tag{3.55}$$

Alternatively, power and torque nominal propeller torque coefficients, K_{pow} and K_Q respectively, can be estimated without curve-fitting using the MCR values for power, torque and rpm, $P_{\text{brake},MCR}$, $Q_{E,MCR}$ and $N_{E,MCR}$. Indeed:

$$K_{pow} = \frac{P_{\text{brake},MCR}}{N_{E,MCR}^{\;3}} = 0.04936 \text{ kW/rpm}^3 \text{ and } K_Q = \frac{Q_{E,MCR}}{N_{E,MCR}^{\;2}} = 0.4753 \text{ kN m/rpm}^2$$

It can be easily seen that the values for the power and torque propeller law coefficients are approximately 10% larger than the ones obtained using curve-fitting. The ones closer to the actual values are, however, the smaller ones. Indeed, in actual shipboard installations, there is a shaft generator present between the main engine and the propeller, for efficient electric power generation. The shaft generator is usually coupled to the ship's electrical system at high main engine loads (close to MCR). Therefore, a significant portion of MCR power is not delivered to the propeller but fed to the vessel's grid. By calculating the power difference at MCR, an estimate of the shaft generator power capacity can be obtained:

$$(K_{pow,(b)} - K_{pow,(a)}) \cdot N_{E,MCR}^{\;3} = (0.04936\text{–}0.04276) \cdot 94^3 = 5.4819 \text{ MW}$$

Of course, the above is only a rough estimate. More accurate values for the torque coefficients K_{pow} and K_Q, as well as shaft generator power capacity, can be obtained if the propeller geometrical features are known. Then, a hydrodynamic calculation technique can be used for accurate determination of propeller power and torque absorption capacity at the various steady-state operating points of the marine plant.

For determining the thermodynamic gain C of the engine the manufacturer's performance curves have been used. For the MAN B&W 9K90MC-VI marine engine the steady-state brake power and torque curves *vs* rated fuel index are given in Figure 3.7.

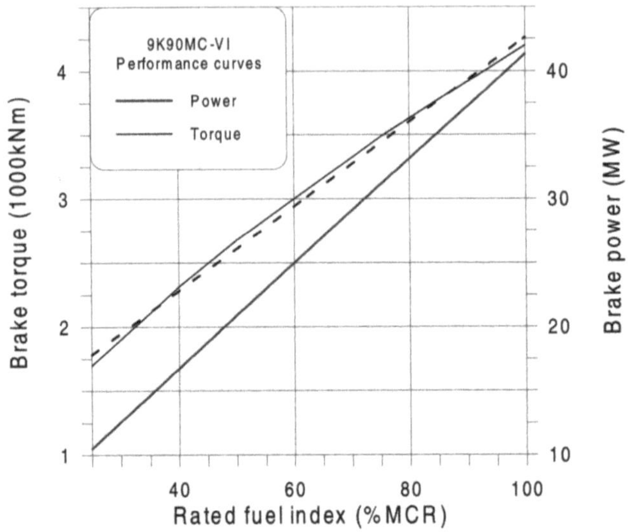

Figure 3.7 "Shanghai Express" plant steady-state curves with respect to fuel index

In Figure 3.7 it is obvious that steady-state engine brake power $P_{\text{brake,SS}}$ is a linear function of steady-state rated fuel index position $F_{R,\text{SS}}$. Indeed, if least-squares curve-fitting (or simply by two steady-state points) is used, the following linear relationship is obtained:

$$P_{\text{brake,SS}} = \kappa_{\text{brake},1} \cdot F_{R,\text{SS}} + \kappa_{\text{brake},0} = 0.41011\ (\text{MW/\%index}) \cdot F_{R,\text{SS}} + 0.33624\ \text{MW}$$

On the other hand, engine steady-state brake torque $Q_{E,\text{SS}}$ is clearly not linear with respect to the corresponding rated fuel index for to the reasons explained earlier. However, by applying least-squares curve-fitting the dashed straight line is obtained with equation:

$$Q_{E,\text{SS}} = 33.05\ (\text{kN m/\%index}) \cdot F_{R,\text{SS}} + 956.907\ \text{kN m}$$

Therefore, as explained earlier, a first estimate for the thermodynamic gain is 33.05 kN m/%index.

In order to obtain a more accurate value for this parameter, especially near MCR, the detailed thermodynamic filling-and-emptying simulation code MoTher was used. Specifically, with the starting point at MCR, step reductions of the engine fuel index by 10% and 20% were imposed. As gain C has to be calculated for engine rotational speed (rpm) maintained as constant as possible, the inertia of the crankshaft has been increased by two orders of magnitude. Therefore, although load torque was kept constant (no propeller loading applied), the resulting engine acceleration was so slow that the effect of the step index reduction on crankshaft rotational speed was completely negligible.

Note that the above procedure, although it reveals the dynamical behaviour of the engine in an undisputed manner and allows determination of all parameters required for control purposes, is infeasible in actual ship propulsion plants, as:

- the load torque applied to the engine is governed by propeller law and cannot be changed in the stepwise manner described above, and,
- the inertia of the plant cannot be increased, so as to allow increase or decrease of fuel index without affecting engine speed.

Even in specialised engine testing facilities the above objectives are rather difficult to implement. Therefore, the use of detailed thermodynamic simulation models, as a substitute of the actual engine, facilitates greatly the propulsion plant modelling for control. Both the above limitations can be surpassed in a numerical simulation model, where both the shaft inertia and load torque can be set as required by the user. Additionally, the turbocharging time constant can be determined with increased accuracy when data from thermodynamic simulation codes are available, instead of experimental series. Indeed, the simulation output series are usually rather pure compared to raw measured series and they also provide increased temporal resolution.

Figure 3.8 shows the engine response, in terms of generated torque, obtained from MoTher code and a first-order transfer function, implemented in Matlab/Simulink, approximating the dynamical behaviour of the engine power/torque generation process.

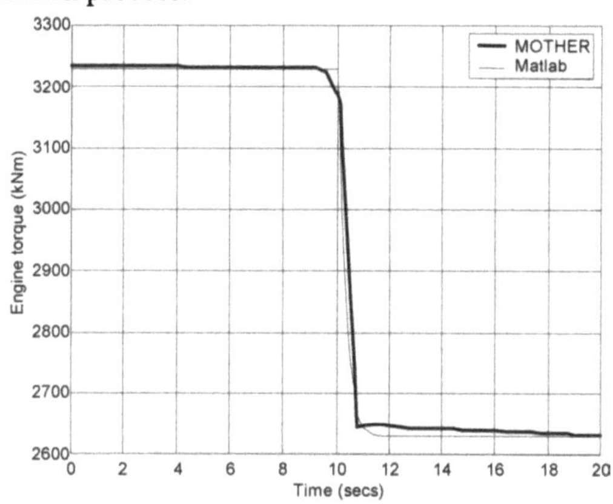

Figure 3.8 Engine torque curves from thermodynamic and transfer function models

The approximating transfer function was of the form:

$$Q_E(s) = \frac{C}{\tau_{TC}s+1} \cdot f_R(s) + Q_{E0} \tag{3.56}$$

The above responses correspond to a 20% fuel index step reduction with infinite (very large) shaft inertia, so that rpm remains practically unchanged, and constant load torque (no propeller loading). Similar index-step simulated experiments were performed with MoTher for 5, 10 and 15% rated fuel index reductions. The conclusions drawn were as follows.

- The value for the thermodynamic gain C calculated in all four cases was found to be 29.0 kN m/%index with remarkable agreement. This supports the assumption that both effective work per cylinder and per cycle, W_e, as well as engine brake torque Q_E, are a linear function of fuel index position F_R, provided that engine speed (rpm) N_E is maintained constant.
- By trial-and-error, the turbocharging/combustion time constant τ_{TC} was determined to be approximately 0.25 s. Although formal identification methods could have been employed, based on least-squares approximation of the step response of the actual process (MoTher) and the first-order transfer function, manual calibration of the transfer function gave satisfactory results, as shown in the comparative response plot above.

Significant difficulties are encountered when attempting to assess the value of the turbocharging time constant τ_{TC} from recorded experimental or service data series. Except for practical sampling problems, the major difficulty arises due to the rpm dependence of propeller-law loading. Indeed, as explained earlier, propeller loading introduces a slow dominant pole at $-\tau_{prop}^{-1} = (-2K_{Q0}N_0 / I)$. This pole cancels, to a large extent, the effect of the fast pole due to turbocharging time constant τ_{TC} on the shaft speed (rpm) signal. Note that in marine practice the engine torque signal is, in most cases, not directly measurable. In contrast, when thermodynamic simulation is employed the engine torque can be obtained, allowing for accurate identification of the turbocharging/combustion dynamical term.

In order to demonstrate the difficulties of estimation for τ_{TC} from actual marine plants, rpm responses of the full-order and reduced-order transfer functions of the marine plant for a 20% fuel index reduction and propeller-law load, are given in Figure 3.9.

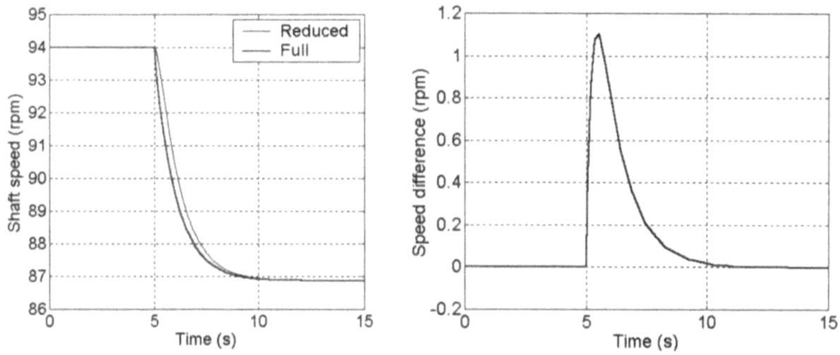

Figure 3.9 Rpm response from the full and reduced-order transfer functions to index step

The step in fuel index occurs at $t = 5$ s. Note that the engine must be ungoverned (manual command of fuel rack) in order to perform the fuel index step. In marine practice, the manual mode of fuel index control is avoided for safety reasons. Therefore, in order to obtain the step speed response of a marine plant, experiments in addition to the vessel's normal operating schedule have to be carried out, as service data series of shaft rpm would include the effect of the speed governor in the dynamic behaviour. In the case where the above-described step experiment cannot be realised, knowledge of the speed governor transfer function is necessary in order to "deduce" its result on the dynamical features.

In any case, as can be seen from the simulation results, the rpm response of the full-order transfer function is quite close to that of the reduced-order one. This is due to the fact that the propeller law introduces a dominant dynamical term, which reduces greatly the effect of the fast turbocharging/combustion pole. Indeed, the value of the turbocharging time constant is (approximately) 0.25 s, whilst the propeller-loading time constant is:

$$\tau_{prop} = \frac{I}{2K_{Q0}N_0} = \frac{\frac{2\pi}{60}\cdot 777350.0\ (\text{kg m}^2\cdot\frac{\text{rad/s}}{\text{rpm}})}{2\cdot 0.44396\cdot 10^3\,\text{N m/rpm}^2\cdot 94.0\ \text{rpm}} = 0.9753\,\text{s} \approx 4\cdot\tau_{TC} = 4\cdot 0.25\,\text{s}$$

From the above it becomes clear that the determination of the turbocharging time constant becomes problematic in the presence of propeller loading, *i.e.* from shipboard shaft rpm series.

Finally, detailed thermodynamic models can be used for validation of prediction by simplified transfer function or state-space models. Therefore, in order to verify that the powerplant behaviour can be accurately predicted using the transfer function model, simulation runs were performed, using both the reduced-order transfer function model implemented in Matlab/Simulink, as well as the MoTher simulation code. For these runs, low-frequency, sinusoidal propeller load profiles were considered. Figure 3.10 shows the response to a sinusoidal disturbance with frequency f = 0.1 Hz and amplitude 300 kN m.

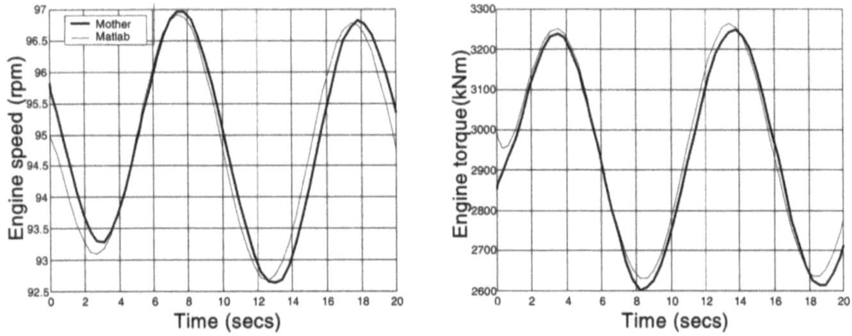

Figure 3.10 Sinusoidal response of the thermodynamic and transfer function models

From Figure 3.10 plot it can be seen that there is satisfactory agreement between the response obtained from the reduced-order transfer function and the one using MoTher. Thus, the reduced-order transfer function can be safely used for further engine control developments.

Finally, note that for the above simulated experiments the engine was governed by a "reference" governor, in contrast to the case of the fuel index step experiment where no feedback control was applied. The term "reference" means that the P and I governor constants, K_P and K_I respectively, which were used in both the transfer function model and in MoTher, were determined in order to give the best agreement between MoTher simulation results and shipboard measured data. The tuning of the governor element in order that the governed engine response matches the shipboard series was manual, and further details are provided in [27]. The PI constant values assumed to designate this reference governor (Ref PI) were:

$$\text{Ref PI: } K_P = 5\%\text{index/rpm and } K_I = 1\%\text{index/(rpm} \cdot \text{s)} \tag{3.57}$$

The block diagram of the closed-loop propulsion system the containership "Shanghai Express" is shown in Figure 3.11.

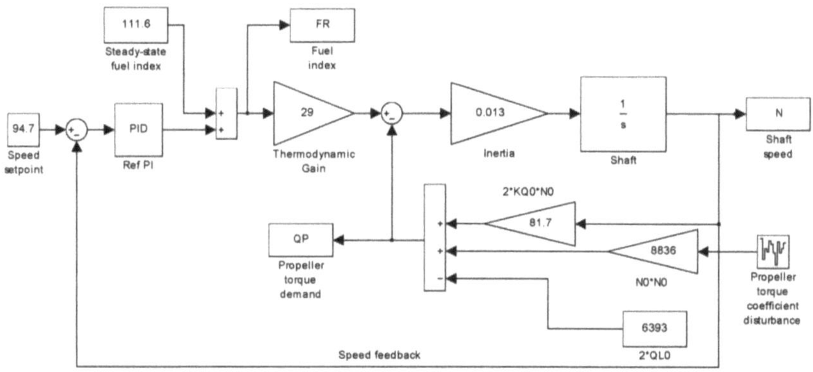

Figure 3.11 The reduced-order transfer function of "Shanghai Express" powerplant

The schematic in Figure 3.11 includes the numerical values of each block participating in the reduced-order transfer function of the "Shanghai Express" propulsion plant. Additionally, the propeller law has been linearised around MCR. Therefore, the schematic in Figure 3.11 corresponds to the transfer function of the propulsion powerplant around MCR. As already explained, the marine plant is quite sensitive to propeller torque coefficient disturbance in the region near MCR. Therefore, it is in the vicinity of MCR where the control and regulation problem obtains its most difficult form.

3.4 Summary

An empirical transfer function, proposed in the literature for depicting the dynamical behaviour of the marine plant, is extended here in order to include the effect of turbocharging and severe propeller torque fluctuation. The extension proposed is based on the physical model and engine simulations performed in the previous chapter, where it was seen that, for rapid transient operation, a fast pole, due to turbocharging and combustion dynamics, needs to be included in the localised transfer function. On the other hand, the marine shafting system dynamical analysis, based on linear systems theory, justified the assumption of ignoring these dynamical terms in speed governor design, provided that certain precautions are taken in order to avoid resonance. Finally, in accordance with practice, as marine engine speed governors are tuned for calm water and/or normal operating mode, a reduced-order transfer function is deduced from the full-order one, where fast dynamical terms are omitted in order to simplify design. For both reduced and full-order transfer functions, an identification procedure is proposed that combines steady-state manufacturer performance data and engine response predicted from detailed thermodynamic simulation models.

CHAPTER 4

ROBUST PID CONTROL OF THE MARINE PLANT

4.1 Introduction

4.1.1 The PID Control Law

Since the early days of feedback control the control scheme with the widest acceptance in industry is the Proportional, Integral and Derivative (PID) controller. Any PID controller implements the PID control law. As indicated by its name, this control law generates a driving signal (control action) based on the values of the tracking error $e(t)$, time integral of $e(t)$ and, possibly, time derivative of $e(t)$. Tracking error $e(t)$ is defined as the deviation of the controlled output's actual value y from its desired setpoint (reference) value y_{set}, at any time instant *t*, *i.e.*:

$$e(t) = y_{set}(t) - y(t) \tag{4.1}$$

Based on the past and present values of tracking error $e(t)$, a (linear) PID controller generates a control action $u(t)$ given by the following relation [33]:

$$u(t) = K_P \cdot e(t) + K_I \cdot \int_{t_0}^{t} e(\xi) d\xi + K_D \cdot \frac{d}{dt} e(t) \tag{4.2}$$

where constants K_P, K_I and K_D are the proportional, integral and differential gains of the control law. On account of the existence of three terms in the equation for the control action signal $u(t)$, PID control is referred to as "three-terms control".

The Laplace transform of the above relation of the PID control law is given by the relation:

$$u(s) = \left(K_P + \frac{K_I}{s} + K_D s \right) \cdot e(s) \tag{4.3}$$

The block diagram of the PID control law is shown in Figure 4.1.

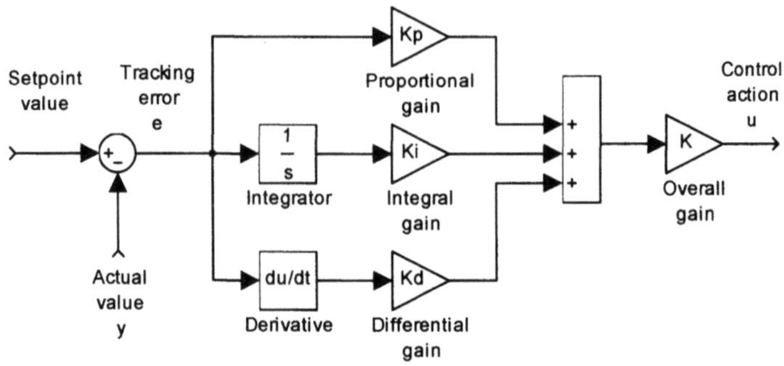

Figure 4.1 Block diagram of the PID control law

Quite often, in control engineering literature, *e.g.* in [24,25], the following slightly modified equation of the PID control law is encountered:

$$u(t) = K \cdot \left(e(t) + \frac{1}{T_i} \cdot \int_{t_0}^{t} e(\xi)d\xi + T_d \cdot \frac{d}{dt} e(t) \right) \tag{4.4}$$

This form of the PID control law is usually referred to as parallel implementation or architecture. It is evident that the proportional constant K is now affecting all terms in the control action. In the rest of this text, however, the forms of Equations 4.2 and 4.3 will be used instead, because they are considered as more straightforward and easier to manipulate. Finally, note that the correspondence between the two forms of the PID control law is obvious and therefore omitted.

The design problem in the case of the PID control law is to determine the three gains of the controller so as certain specifications in the time and or frequency domain are met. This is referred to by the term PID controller tuning, and will be examined closer for the case of marine plants later in this chapter. Although, in theory, all possible combinations of the three terms of the PID control law are possible, in engineering practice only P, PI and PID controllers are encountered. Therefore, a brief analysis of these types is given in the rest of this section.

4.1.2 Proportional Control

In proportional control (P-control): $K_I = K_D = 0$. Therefore, the tuning problem is reduced to the determination of K_P. Historically, P-control has emerged as an extension to on/off control and mainly for avoiding the effect of "ringing" (oscillation due to continuous activation and deactivation of the control action). In practice, the output of a feedback P-controller is superimposed on that of a feedforward (open-loop) controller. The overall control action $u(t)$ is calculated as:

$$u(t)=\begin{cases}u_{max}, & e(t)>e_0\\ u_0+K_P\cdot e(t), & -e_0\le e(t)\le e_0\\ u_{min}, & e(t)<-e_0\end{cases}\qquad(4.5)$$

Note that, in practical applications, the proportional control action is saturated when the output error $e(t)$ is outside an interval around zero. This is done in order to avoid excessive values and related instabilities of a non-linear nature.

The major disadvantage of P-control is that, in most cases, steady-state error is generated. Specifically, if the simple case at which steady-state process output y is proportionally dependent (through a process steady-state gain K_{SS}) on the control action u is considered:

$$y=K_{SS}\cdot u=K_{SS}\cdot(u_0+K_P\cdot e)\qquad(4.6)$$

This is in most cases (at least approximately) true for steady-state analysis around an operating point. Assuming P-control and by solving the above with respect to y we obtain that:

$$y=\frac{K_{SS}\cdot u_0+K_{SS}K_P\cdot y_{set}}{1+K_{SS}K_P}\Rightarrow e=\frac{y_{set}-K_{SS}\cdot u_0}{1+K_{SS}K_P}\qquad(4.7)$$

Therefore, in order to minimise (ideally eliminate) steady-state error, at least one of the following must hold:

- $u_0=y_{set}/K_{SS}$
- $K_P\to\infty$

For the case of marine plants, as explained in Section 3.3.3, deviation $n(s)$ of shaft rpm from the steady-state value N_0 is given by:

$$n(s)=\frac{C}{Is+2K_{Q0}N_0}\cdot f_R(s)-\frac{N_0^2}{Is+2K_{Q0}N_0}\cdot k_Q(s)\qquad(4.8)$$

Steady-state deviation $n(t\to\infty)$ is calculated for a step change in disturbance k_Q, according to the Final Value Theorem of the Laplace Transform [33], by:

$$n(t\to\infty)=\lim_{s\to 0}\left(s\cdot n(s)\right)=\frac{C}{2K_{Q0}N_0}\cdot f_R(t\to\infty)-\frac{N_0}{2K_{Q0}}\cdot k_Q(t\to\infty)\qquad(4.9)$$

By assuming P-control:

$$f_R=K_P\cdot n\qquad(4.10)$$

Therefore:

$$n(t\to\infty)=\frac{N_0^2}{CK_P-2K_{Q0}N_0}\cdot k_Q(t\to\infty)\qquad(4.11)$$

In order for steady-state error (deviation) $n(t\to\infty)$ to be small, it is required either to introduce a large value for the proportional gain K_P or to appropriately match the steady-state fuel index position F_{R0} to the propulsion power/torque requirements. Indeed, according to the considerations explained in Chapter 3:

$$C \cdot F_{R0} - Q_{fr} = K_{Q0} \cdot N_0^2 \Rightarrow k_Q(t \to \infty) = 0 \Rightarrow n(t \to \infty) = 0 \tag{4.12}$$

Before concluding the section for P-control, an implementation of the P-controller, using simple, and low-cost analogue electronic components is shown in Figure 4.2.

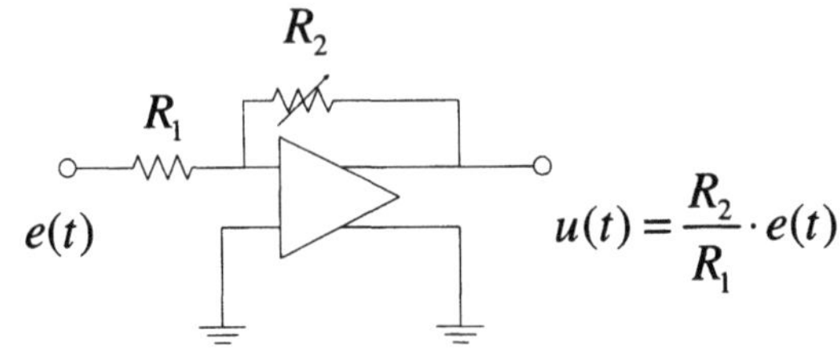

Figure 4.2 Analogue circuit implementation of the P-controller

This implementation of the P-controller is comprised of two resistors (one adjustable) and an operational amplifier (op-amp). Today, however, analogue electronics are employed for controller implementation rather seldom. On the other hand, digital implementations using microcontrollers and/or Programmable Logic Controllers (PLC) dominate in maritime and industrial applications. Detailed analysis of PID control law digital implementations, based on the Z-transform instead of the Laplace transform, can be found in literature, *e.g.* [9,10].

4.1.3 Proportional-Integral Control

In proportional-integral control (PI-control): $K_D = 0$. This type of control has been introduced in order to overcome the deficiency of P-control to compensate steady-state errors. Due exactly to this specific feature, PI-controllers form the vast majority of industrial and process controllers [23]. In fact, most of today's marine engine speed regulators (governors) are of the PI-controller type [34]. Furthermore, it can be seen that flyweight (Woodward) mechanical governors can be represented as a system with a transfer function analogous to the generic transfer function of the PI-controller.

The PI-controller can eliminate steady-state errors, by introducing a zero at $s = 0$. This effect will be demonstrated in the specific case of the reduced-order transfer function of the marine plant. However, it is valid for any first-order transfer function of the following form, as introduced in the previous chapter:

$$H(s) = \frac{y(s)}{u(s)} = K \cdot \frac{e^{-\tau_D s}}{\tau s + 1} \tag{4.13}$$

A PI-controller generates a control action $u \equiv f_R$, which takes the following form in the case of the marine propulsion system:

$$f_R(t) = K_P \cdot n(t) + K_I \cdot \int_{t_0}^{t} n(\xi) d\xi$$

$$\Updownarrow$$

$$f_R(s) = \left(K_P + \frac{K_I}{s} \right) \cdot n(s) = \frac{K_P \cdot s + K_I}{s} \cdot n(s) \tag{4.14}$$

The closed-loop transfer function is determined by substituting the above expression for the control action $f_R(s)$ in the reduced-order open-loop transfer function. In effect, we obtain that:

$$n(s) = \frac{C}{Is + 2K_{Q0}N_0} \cdot \frac{K_P \cdot s + K_I}{s} \cdot n(s) - \frac{N_0^2}{Is + 2K_{Q0}N_0} \cdot k_Q(s)$$

$$\Downarrow$$

$$n(s) = -\frac{N_0^2 \cdot s}{I \cdot s^2 + (2K_{Q0}N_0 - CK_P) \cdot s - CK_I} \cdot k_Q(s) \tag{4.15}$$

Therefore, steady-state error $n(t \to \infty)$ (*i.e.* deviation from steady-state shaft rpm value N_0 for a step change in disturbance k_Q) is driven to zero. Indeed, using the Final Value Theorem of the Laplace Transform:

$$n(t \to \infty) = \lim_{s \to 0} \left(s \cdot n(s) \right) = -\frac{N_0^2 \cdot 0}{I \cdot 0^2 + (2K_{Q0}N_0 - CK_P) \cdot 0 - CK_I} \cdot k_Q(t \to \infty) = 0 \tag{4.16}$$

Note that the above holds provided that $K_I \neq 0$. Indeed, if this gain is set to zero, then zero-pole cancellation occurs in the above closed-loop transfer function and the control scheme employed "falls back" to P-control. Therefore, the utilisation of rpm time integral feedback allows damping of steady-state error, introduced by a possible disturbance $k_Q(t \to \infty) \neq 0$, *i.e.* a value for steady-state torque coefficient K_Q other than the nominal one K_{Q0}.

Another prerequisite for elimination of steady-state error is that both poles of the closed-loop transfer function are stable. This can be guaranteed, if and only if, the proportional and integral gains are selected so that both roots of the denominator polynomial lie in the left half s-plane (complex plane).

However, zero steady-state error is not achieved without a penalty. Indeed, as can be seen, the order of the closed-loop transfer function is increased by one. Therefore, the response may become oscillatory, as the second-order denominator polynomial may obtain roots with nonzero imaginary parts, although its coefficients are real. On the other hand, P-control, which can be regarded as a sub-case of PI-control when integral gain is zeroed $(K_I = 0)$, cannot give an oscillatory response. Indeed, the denominator polynomial becomes in the case of P-control:

$$p_c(s) = I \cdot s + (2K_{Q0}N_0 - CK_P) \Rightarrow s_0 = -\frac{2K_{Q0}N_0 - CK_P}{I} \tag{4.17}$$

Therefore, the denominator polynomial obtains a single real pole. Provided that:

$$K_P < 2K_{Q0}N_0 / C \Rightarrow s_0 < 0 \tag{4.18}$$

the system is stable and the response demonstrates no oscillation.

Another quite interesting sub-case is the one when the denominator polynomial has a zero first-order term. This can happen if:

$$K_P = 2K_{Q0}N_0 / C \tag{4.19}$$

In that case the denominator polynomial becomes:

$$p_c(s) = I \cdot s^2 + 0 \cdot s - CK_I \Rightarrow s_{1,2} = \pm\sqrt{\frac{CK_I}{I}} \tag{4.20}$$

Therefore, the closed-loop system is stable (but not asymptotically stable, *i.e.* the response is an undamped sinusoidal) if and only if:

$$K_I < 0 \Rightarrow s_{1,2} = \pm j\sqrt{\frac{C|K_I|}{I}} \tag{4.21}$$

In the opposite case, *i.e.* when $K_I \geq 0$, the closed-loop system is unstable because one of its poles is located in the right *s*-plane. Therefore, the role of the P-term in the PI-controller is to provide closed-loop stability; as demonstrated, a specific value, such as the one zeroing the first-order term, can prevent the closed-loop system from being asymptotically stable.

The physical interpretation of the fact that the PI-controller eliminates steady-state error is that when this type of control is employed the transient is not terminated until the controlled output has reached its setpoint (reference) value. This is due to the accumulation of error in the I-term, which, therefore, comprises a form of system memory. In that respect, the control action is driven not only by the current value of tracking error, but also by the history of the error. In conclusion, the transient is not terminated until the integral error is zeroed. In order, to achieve this, however, the controller may drive the controlled output to oscillate so that the time integral of the error is compensated.

Practical implementation of the I-term involved in the PI-controller, by using analogue electronic components is demonstrated in the following Figure 4.3.

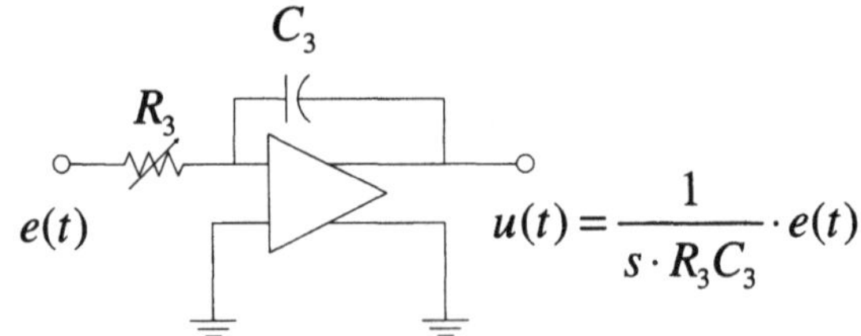

Figure 4.3 Analogue circuit implementation of the I-term for PI control

This low-cost implementation includes an op-amp, to the inverting input of which a capacitor and an adjustable resistor are connected. The integration time constant $T_i = R_3C_3$ can be modified through the value of the resistor.

A full implementation of the PI-controller can be effectuated by combining a P-controller circuit, as presented previously, with an I-term circuit. This is achieved by the use of an analogue summation circuit. This looks exactly like the P-controller. At the inverting end of the op-amp all the signals to be added are connected. The intermittence of the op-amps simplifies the design and enhances the circuit insensitivity. However, a detailed analysis of analogue controller implementations is outside the scope of this work and therefore will be omitted.

Finally, note that, as in the case of the P-controllers, digital implementations using microcontrollers and/or Programmable Logic Controllers (PLC) are preferable nowadays. Detailed analysis of the PI-controller digital implementation, based on the *Z*-transform, can be found in literature, *e.g.* [9,10].

4.1.4 Proportional-Integral-Derivative Control

This is the full version of the PID control law; none of the three terms involved is zero. PID-control has evolved from the PI-control. The practical need for the introduction of the D-term in the control law is reduction of overshoot due to the I-term. As already mentioned, a nonzero I-term may give rise to undesired oscillation in the controlled output. Indeed, because the order of the transfer function is increased by one, the poles of the closed-loop system may obtain a nonzero imaginary part. Therefore, an oscillation is introduced, which results in overshoot. Figure 4.4 shows a typical 0–100% step response at $t = 0$ s of a closed-loop system with PI-control is plotted.

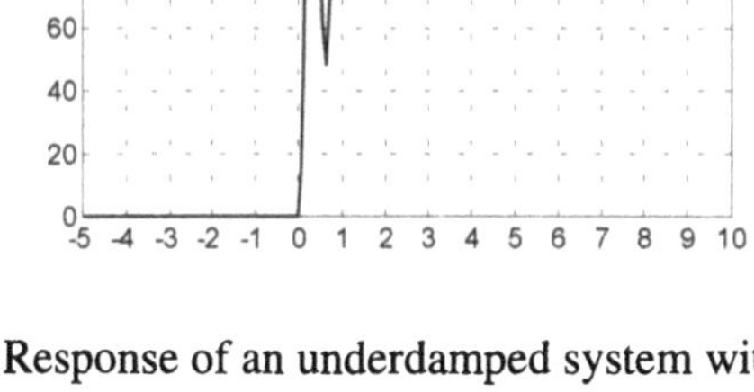

Figure 4.4 Response of an underdamped system with PI control

It is obvious that excessive overshoot (approximately 70% of the steady-state value) is introduced as a side-effect of the I-term employment. A method to decrease such prohibitive values of overshoot is to use a nonzero D-term in the control law.

Introduction of the D-term in the transfer function of the marine plant is demonstrated hereafter. The PID-controller generates a control action according to the following equation:

$$f_R(s)=\left(K_P+\frac{K_I}{s}+K_D s\right)\cdot n(s)=\frac{K_I+K_P s+K_D s^2}{s}\cdot n(s) \tag{4.22}$$

Therefore, the closed-loop transfer function from the disturbance signal $k_Q(t)$ to the speed deviation signal $n(t)$ is as follows:

$$n(s)=\frac{C}{Is+2K_{Q0}N_0}\cdot\frac{K_I+K_P\cdot s+K_D\cdot s^2}{s}\cdot n(s)-\frac{N_0^2}{Is+2K_{Q0}N_0}\cdot k_Q(s)$$

$$\Downarrow$$

$$n(s)=-\frac{N_0^2\cdot s}{(I-CK_D)\cdot s^2+(2K_{Q0}N_0-CK_P)\cdot s-CK_I}\cdot k_Q(s) \tag{4.23}$$

It is evident that all coefficients in the denominator polynomial of the above closed-loop transfer function are adjustable, through one of the three gains of the PID-controller. Therefore, the roots can be located as desired, so that:

- oscillation is reduced,
- excessive overshoot is drastically reduced, too, or even completely eliminated.

In parallel, however, the transient performance of the closed-loop system can be maintained at the same levels, in terms of speed of response and relative stability.

An interesting sub-case of the PID-controller, is the PD-controller, *i.e.* when integral gain is set to zero $(K_I=0)$. Then the closed-loop transfer function becomes:

$$n(s)=-\frac{N_0^2}{(I-CK_D)\cdot s+(2K_{Q0}N_0-CK_P)}\cdot k_Q(s) \tag{4.24}$$

As in the case of P-control, PD-control cannot suppress steady-state error. Indeed, by applying the Final Value Theorem to the above, the following is obtained for a step change in disturbance k_Q:

$$n(t\to\infty)=\frac{N_0^2}{CK_P-2K_{Q0}N_0}\cdot k_Q(t\to\infty) \tag{4.25}$$

Therefore, in the case of marine plants PD-control gives identical results with P-control as far as steady-state error is concerned. However, in [33] it is argued that PD-control can affect steady-state error for systems whose steady-state error changes with time, as derivative de/dt is nonzero. In any case, any transient specification met by PD-control is achievable with P-control, too. Indeed, the single, real, closed-loop pole obtained with PD-control is:

$$s_0 = -\frac{2K_{Q0}N_0 - CK_P}{I - CK_D} \tag{4.26}$$

An "equivalent" P-controller, with gain $K_{P,\mathrm{EQ}}$ can be used in order to obtain the same pole. By equating the above with the expression for the pole of the marine plant when only P-control is employed:

$$\frac{2K_{Q0}N_0 - CK_{P,\mathrm{EQ}}}{I} = \frac{2K_{Q0}N_0 - CK_P}{I - CK_D}$$

$$\Downarrow$$

$$K_{P,\mathrm{EQ}} = \frac{I}{I - CK_D} \cdot K_P - \frac{2K_{Q0}N_0}{I - CK_D} \cdot K_D \tag{4.27}$$

However, if the pole assigned to the system with P-control is identical to the pole of the system with PD-control, then the steady-state error of P-control is different. Indeed, remember that the steady-state error of the P-controlled system is:

$$n(t \to \infty) = \frac{N_0^2}{CK_{P,\mathrm{EQ}} - 2K_{Q0}N_0} \cdot k_Q(t \to \infty) \tag{4.28}$$

It is evident that use of PD-control of marine plants allows meeting a combined transient, as well as steady-state error specifications. On the other hand, P-control cannot provide this flexibility; specifications should be restricted to transient performance or steady-state error. Finally, PD-control can achieve zero steady-state error only if $K_P \to \infty$. Therefore, only the full PID-control law can provide zero steady-state error, combined with desired transient performance, but with the penalty of increasing the order of the closed-loop dynamics.

Introduction of the D-term in the control law gives an "anticipatory" ability to the controller, since derivative de/dt is the slope of $e(t)$. Normally, in linear or linearised systems, if the slope of $e(t)$ is large, *e.g.* due to a step input, a high overshoot will subsequently occur. The derivative control measures the instantaneous (actually from one thermodynamic cycle to the next in the case of reciprocating engines) slope of $e(t)$, predicts the large overshoot ahead of time, and makes a proportional correcting effort before the overshoot actually occurs.

The D-term in PID-control can be implemented by the analogue electronic circuit shown in Figure 4.5.

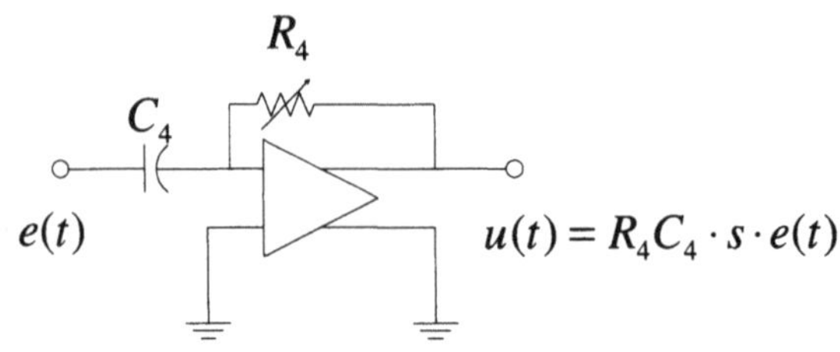

Figure 4.5 Analogue circuit implementation of D-term for PID control

Evidently, the differentiation time constant $T_d = R_4 C_4$ can be modified through the value of the adjustable resistor. Here, it should be noted that the practical difficulty with any implementation of a controller including a D-term is that the differentiator behaves as high-pass filter, which allows high-frequency (HF) noise to propagate through the closed-loop system. Major sources of HF noise are the measuring devices, which introduce the well-known measurement error. Although advanced filtering techniques have been developed, aiming to reduce the effect of measurement error, it is still advisable to avoid differentiation of measured signals. Measurement error is the major reason for which practical controllers rarely employ D-term based action. However, in the context of the present work a method is given in order to bypass this deficiency of direct D-term implementations, which otherwise is most valuable for improvements in transient performance. The idea is to use additional feedback signals, commonly available in modern marine propulsion powerplants, in order to estimate the derivative of shaft rpm without the need to differentiate the measured shaft rpm feedback signal itself.

4.2 Application Aspects of Marine Engine Governing

4.2.1 Functionality Requirements

The PID control law is the principal linear module operating in the core of any marine engine speed governor, employed shipboard for rpm regulation. Of course, the core PID control law is surrounded by a number of functional modules ensuring safe powerplant operation and remote control facilities. Such modules include the bridge/engine room telegraph, allowing transfer of engine command to the bridge, as well as interconnection to ship-wide data acquisition (DAQ), monitoring/alarm and fault-diagnostic networks.

Other functional modules of the marine propulsion engine governor ensure safe powerplant operation. These modules are actually slave control loops for various variables of the propulsion plant. For example, they may include engine coolant (water) temperature and pressure, intercooler coolant temperature, lubrication oil pressure and temperature, scavenging air pressure, exhaust gas temperature, *etc*. The form of control imposed is most often of the on/off type and used for:

- Switching on and off (driving) auxiliary machinery such as additional heat exchangers for coolants or feed pumps for lubrication oil.
- Applying a limit (usually on the upper side) on engine loading, as expressed by the value of rated fuel index, so that operational safety is maintained and possible engine failure is prevented.

A related functional module is the one ensuring that transition of engine rpm through a number of barred regions of the operating range is effectuated rapidly. Indeed, such rpm regions interrupt the continuity of the allowed operating range of the engine. The cause of this is that, in the specified regions, the discrete cylinder firings give an excitation whose frequency lies in the close vicinity of the shafting system resonance points.

Another functional module of modern marine engine governors is starting air control. Pressurised air is supplied to the engine cylinders when the shaft is still, in order to initiate rotation and, in effect, compression in the combustion chambers. Then, if fuel is supplied, combustion may get started and the engine can accelerate in the normal or reverse sense of rotation. Starting air control includes the coordination of all actions for the engine to start running in normal or reverse mode, such as: sensing of shaft stillness, starting air transmission control with proper cylinder sequencing for normal or reverse rotation, fuel injection command, *etc*. The resulting algorithm can grow become rather complicated and is by itself a subject of a separate study and analysis.

In the framework of the present work the major subject of interest will be rpm and operating point regulation of the marine powerplant. Thus, in the rest of this chapter PID controller tuning methods are examined and developed. First, however, the nature of engine torque generation and propeller torque absorption are investigated in the next paragraph, so that the requirements for the speed governor can be clearly understood.

4.2.2 Spectral Analysis of Engine and Propeller Torque

As demonstrated in the current paragraph, the discrete nature of engine torque delivery, in combination with the finite number of propeller blades, limits the potential of the engine speed governor's rpm regulating function to the speed variation from one engine cycle to another.

Focusing the analysis on the steady-state, and assuming that the cycle-mean shaft speed is N, expressed in rpm, engine and propeller torque power spectral density (psd) functions, $|Q_E(\omega)|^2$ and $|Q_P(\omega)|^2$ respectively, and, in effect, the magnitude of the Fourier transforms (magnitude spectra), $|Q_E(\omega)|$ and $|Q_P(\omega)|$ respectively, can be analysed to the following frequency components:

$$|Q_E(\omega)| = Q_{E,\mathrm{LF}}(\omega, N) + Q_{E,\mathrm{HF}}(\omega, z_c \cdot N)$$
$$\text{and} \tag{4.29}$$
$$|Q_P(\omega)| = Q_{P,\mathrm{LF}}(\omega, N) + Q_{P,\mathrm{HF}}(\omega, \mathrm{N_{blades}} \cdot N)$$

In the above $\mathrm{N_{blades}}$ is the number of propeller blades and z_c is the number of engine cylinders. The low-frequency component in both the above magnitude spectra has the following property:

$$\int_0^{+\infty} Q_{E \text{ or } P,\mathrm{LF}}(\omega, N) \cdot d\omega \approx \int_0^{2\pi N/60} Q_{E \text{ or } P,\mathrm{LF}}(\omega, N) \cdot d\omega \tag{4.30}$$

Moreover, steady-state engine-load matching is translated to:

$$Q_{E,\mathrm{LF}}(\omega = 0) = Q_{P,\mathrm{LF}}(\omega = 0) \tag{4.31}$$

otherwise, cycle-mean shaft speed cannot be maintained constant. For values of ω between 0 and $2\pi N/60$, $Q_{E,\mathrm{LF}}$ is determined primarily by the fuel index value and $Q_{P,\mathrm{LF}}$ by the sea conditions. Here, remember that fuel index variation is maintained adequately slow so that the in-cycle variation of fuel mass injected in each cylinder is small enough, so that eigenmodes of the shafting system are not excited (see analysis in Section 3.2.1).

On the other hand, the following hold for the high-frequency components of the engine and propeller torque magnitude spectra:

$$\int_0^{+\infty} Q_{E,\mathrm{HF}}(\omega, z_c \cdot N) \cdot d\omega \approx \int_{z_c \cdot 2\pi N/60}^{+\infty} Q_{E,\mathrm{HF}}(\omega, z_c \cdot N) \cdot d\omega \tag{4.32}$$

$$\int_0^{+\infty} Q_{P,\mathrm{HF}}(\omega, \mathrm{N_{blades}} \cdot N) \cdot d\omega \approx \int_{\mathrm{N_{blades}} \cdot 2\pi N/60}^{+\infty} Q_{P,\mathrm{HF}}(\omega, \mathrm{N_{blades}} \cdot N) \cdot d\omega \tag{4.33}$$

The above facts for the torque magnitude spectra, which can be derived from the engine thermodynamics and propeller hydrodynamics [35], also determine the nature of the speed governor intervention for rpm regulation. Specifically, the speed governor can affect, through fuel index (rack) adjustment, $Q_{E,\mathrm{LF}}(\omega > 0)$ in order to reduce the effect on shaft rpm introduced by $Q_{P,\mathrm{LF}}(\omega > 0)$. Also, note that steady-state engine loading $Q_{E,\mathrm{LF}}(\omega = 0) = Q_{E0}$ is determined by the steady-state fuel index value F_{R0}, because, as already analysed in Section 3.3.2, it holds that $Q_{E0} = C \cdot F_{R0} - Q_{fr}$.

Although care is taken that fuel index changes do not excite shafting eigenmodes, it is possible that this happens due to $Q_{E,\mathrm{HF}}(\omega, z_c \cdot N)$. As already mentioned, fuel index controls only the cycle-mean part of engine torque, namely $Q_{E,\mathrm{LF}}(\omega, N)$. On the other hand, $Q_{E,\mathrm{HF}}(\omega, z_c \cdot N)$ exhibits maxima at multiples of the principal harmonic $(z_c \cdot 2\pi N/60)$ due to the discrete cylinder firings. Using the result obtained in Section 3.2.1 for the shafting system resonance frequency f_c, the following estimate for the rpm values at which resonance of the shafting system is dangerous to occur can be obtained:

$$z_c \cdot 2\pi N / 60 = 2\pi f_c \Rightarrow N = \frac{60}{z_c} \cdot \sqrt{\left(\frac{1}{I_E} + \frac{1}{I_P}\right) \cdot K_{shaft}} \approx \frac{120}{z_c} \cdot \sqrt{\frac{K_{shaft}}{I}} \quad (4.34)$$

Many modern marine engine governors include functions for providing an appropriate alarm or even automatically adjusting the rpm setpoint in order to avoid shafting system overloading due to resonance. Specifically, transitions through the dangerous rpm values are rapid and usually one is not allowed to stabilise the plant around such a resonance rpm value [34,35].

Finally, taking into account the above spectral decomposition for engine and propeller torque, it is now argued that a significant amount of noise is introduced in the signal of shaft rpm $N(t)$, measured for control purposes which is the major plant feedback signal to the governor. The noise introduced is due to the high-frequency components of engine and propeller torque, $Q_{E,\mathrm{HF}}(\omega, z_c \cdot N)$ and $Q_{P,\mathrm{HF}}(\omega, \mathrm{N}_{\mathrm{blades}} \cdot N)$ respectively. This type of noise on the rpm signal, in order to be clearly discriminated from other noise types, such as measurement-induced noise, will be referred to by the term "process noise".

The analysis will be limited in the case of MCR operation, without any loss of generality, as it can be easily extended to any other steady-state operating point. In the following, f_{MCR} and ω_{MCR} are defined as:

$$f_{MCR} \text{ (rps or Hz)} = \frac{N_{MCR} \text{ (rpm)}}{60 \text{ s/min}} \text{ and } \omega_{MCR} \text{ (rad/s)} = 2\pi \cdot f_{MCR} = 2\pi \cdot \frac{N_{MCR}}{60} \quad (4.35)$$

The following idealised expressions are considered for $Q_{E,\mathrm{HF}}$ and $Q_{P,\mathrm{HF}}$, where only the first harmonic of these torque spectral components appears:

$$Q_{E,\mathrm{HF}}(t) = Q_{E,\mathrm{HF0}} \cdot \cos(z_c \cdot \omega_{MCR} t) \text{ and } Q_{P,\mathrm{HF}}(t) = Q_{P,\mathrm{HF0}} \cdot \cos(\mathrm{N}_{\mathrm{blades}} \cdot \omega_{MCR} t) \quad (4.36)$$

Using the equation of motion of the engine–propeller shaft, considered as a lumped element, we obtain that:

$$\dot{n}(t) = \frac{Q_E(t) - Q_P(t)}{I} = \frac{(Q_{E,\mathrm{LF}} + Q_{E,\mathrm{HF}}(t)) - (Q_{P,\mathrm{LF}} + Q_{P,\mathrm{HF}}(t))}{I} \quad (4.37)$$

In steady-state the low-frequency torque spectral components can be assumed equal; thus:

$$\begin{aligned} n(t) &= \int_0^t \frac{Q_{E,\mathrm{HF}}(\xi) - Q_{P,\mathrm{HF}}(\xi)}{I} \cdot d\xi \\ &= \frac{Q_{E,\mathrm{HF0}} \cdot \sin(z_c \cdot \omega_{MCR} t) - Q_{P,\mathrm{HF0}} \cdot \sin(\mathrm{N}_{\mathrm{blades}} \cdot \omega_{MCR} t)}{I \cdot 2\pi f_{MCR}} \end{aligned} \quad (4.38)$$

where it has been set that $n(t=0) = 0$. The above expression is a very rough estimate of in-cycle deviation of shaft rpm from the cycle-mean, steady-state value

N_{MCR}. Although this estimate is rather crude and obtained on the basis of a number of assumptions, it can be used for drawing the following major conclusions. (i) The cycle mean value of the in-cycle rpm fluctuation is zero, as assumed without any detailed justification so far. (ii) In-cycle rpm fluctuation exhibits rich spectral content in frequencies that are multiples of the quantity $(2\pi N_0/60)$, where N_0 is the steady-state value of shaft rpm at each operating point; this consists of the process noise referred to earlier. (iii) The level (power) of process noise depends on the magnitude of the in-cycle variation of engine and propeller, as quantified in the above idealised representation by the amplitudes $Q_{E,\mathrm{HF0}}$ and $Q_{P,\mathrm{HF0}}$ of the sine waves; these parameters are expected to increase with engine loading and speed. On the other hand, the integrating action of the shaft introduces an inverse dependence of the noise level on shaft frequency of rotation (*i.e.* rpm). (iv) The process noise introduced in the shaft rpm signal makes the operation of differentiation (used in the D-term of the PID control law) even more problematic, as analysed further on. Also, the use of a low-pass filter, prior to feeding the rpm signal to the governor, is required, even if high-precision measurement (no measurement noise) is employed.

Numerical investigation of the aspects of marine engine governing related to the spectral features of engine and propeller torque, as analysed before, is presented in the following example.

4.2.3 Example of Propulsion Plant Analysis

Section 3.2.2 presented an analysis of the shafting system of the containership "Shanghai Express". Using the two-mass plus spring, lumped-parameters model presented in Section 3.2.1, the critical frequency of the shafting system has been determined as:

$$f_c = 6.05 \text{ Hz}$$

Taking into account that the main engine of "Shanghai Express" is the nine-cylinders MAN-B&W 9K90MC-VI, the rpm value at which resonance of the shafting system may occur is calculated as follows:

$$N = 60 \cdot f_c / z_c \Rightarrow N = 40.3 \text{ rpm}$$

As usual in marine propulsion installations, resonance of the shafting system may occur only in low rpm values corresponding to low engine loading. The value of 40 rpm is therefore rather typical. Operation of the main engine around this rpm value is avoided and transition is effectuated as rapidly as possible.

The process noise appearing in the rpm plant feedback signal due to the discrete cylinder firings and the finite propeller blade number will now be investigated in the case of "Shanghai Express" by use of the simplified expressions for the first harmonic of high-frequency torque components $Q_{E,\mathrm{HF}}(\omega, z_c \cdot N_{MCR})$ and $Q_{P,\mathrm{HF}}(\omega, \mathrm{N_{blades}} \cdot N_{MCR})$. Here it is mentioned that "Shanghai Express" is propelled by a single, fixed-pitch propeller with six blades and diameter of 8.33 m. Also, as calculated in Section 3.2.2, the shafting system inertia is $I = 777350.0\ \text{kg m}^2$. Finally, the amplitudes, $Q_{E,\text{HF0}}$ and $Q_{P,\text{HF0}}$, of the sine waves, $Q_{E,\text{HF}}(t)$ and $Q_{P,\text{HF}}(t)$, forming the idealised time-domain representations of the high-frequency torque components have been assigned the following *ad hoc* values, for the sake of this investigation:

$$Q_{E,\text{HF0}} = \frac{Q_{E,MCR}}{z_c} = \frac{4.1979 \cdot 10^3\ \text{kN m}}{9} = 466.4\ \text{kN m}$$

$$Q_{P,\text{HF0}} = \frac{K_{Q0} \cdot N_{MCR}^2}{2 \cdot \text{N}_{\text{blades}}} = \frac{0.4753 \dfrac{\text{kN m}}{\text{rpm}^2} \cdot (94\ \text{rpm})^2}{2 \cdot 6} = 349.8\ \text{kN m}$$

Figure 4.6 shows plots of the engine torque delivery, propeller load torque, shaft rpm and rpm time derivative for a time interval of two full shaft revolutions; the period of shaft revolution is $1/f_{MCR} = 0.638$ s. Note that steady-state is assumed *i.e.*, $Q_{E,\text{LF}}(\omega, N_{MCR}) = Q_{P,\text{LF}}(\omega, N_{MCR})$, for all ω.

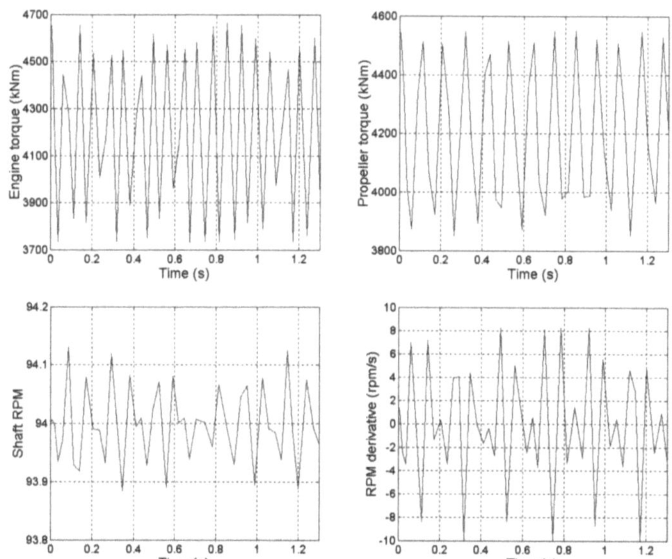

Figure 4.6 Typical shafting system steady-state response

It can be readily concluded that, even in this simplified, idealised situation, process noise is present in the rpm feedback signal. A great deal of the process noise can be removed by use of a low-pass filter in the feedback path of the closed-loop. However, even if filtered, the rpm signal cannot be differentiated; as argued later on, the operation of differentiation reduces to a greater extent the noise rejection effect of any linear low-pass filter. Note here that the in-cycle variation of rpm or rpm derivative is not of value for the governor. Only the cycle-mean values of both these variables are needed. In the case of steady-state, rpm cycle-mean is equal to N_0 and cycle-mean rpm time derivative is zero; therefore, fuel index should be maintained fixed to the value that provides the power required for propulsion, according to engine–propeller matching.

4.3 PID H-infinity Loop Shaping

4.3.1 Theoretical Note

Loop shaping is a major technique applied quite often in control engineering for obtaining the transient performance of a system or process complying with certain requirements; origins of loop shaping can be sought after in pole placement techniques. The idea is rather simple and straightforward. With respect to the block diagram of a Single-Input-Single-Output (SISO) system shown in Figure 4.7

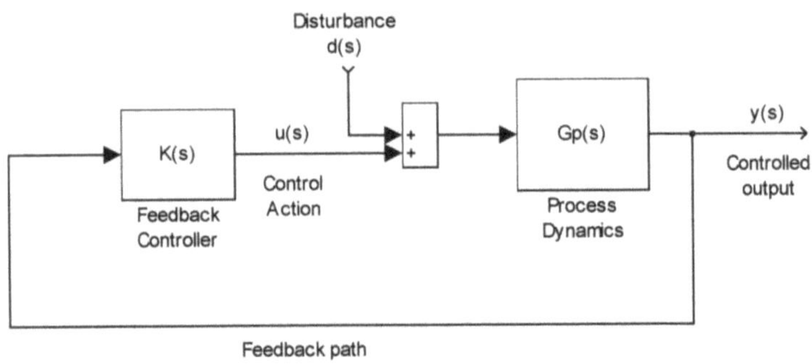

Figure 4.7 Block diagram of compensated SISO system with disturbance

the controller transfer function parameters are selected so that the poles, zeroes and steady-state gain of the closed-loop (scalar) transfer function coincide with the ones required. As can be easily seen, the closed-loop transfer function from (input) disturbance $d(s)$ to the controlled (regulated) output $y(s)$, is:

$$G_c(s) = \frac{G_P(s) \cdot K(s)}{1 + G_P(s) \cdot K(s)} \tag{4.39}$$

Due to controllability and observability considerations, however, formulation of the loop shaping (or in simpler cases pole-placement) control problem is often given in a state-space framework instead of the transfer function approach presented above. This approach is postponed until Chapter 6, when the full-state-feedback control problem of the marine plant is examined. For the rest of this chapter, controllability and observability will be assumed, as, in any case, any zero-pole cancellation manifests loss of system controllability and/or observability.

4.3.2 PID Controller Tuning for Loop Shaping

The PID controller design for the case of marine propulsion powerplants will be based upon the reduced-order transfer function introduced in the previous chapter. The reduced-order transfer function is repeated here for compactness:

$$n(s)=\frac{C}{Is+2K_{Q0}N_0}\cdot f_R(s)-\frac{N_0^2}{Is+2K_{Q0}N_0}\cdot k_Q(s) \tag{4.40}$$

By introducing the full PID control law for closing the control loop of the plant, the following closed-loop transfer function, $G_c(s)$, from the disturbance signal $k_Q(s)$ to the shaft rpm deviation $n(s)$ is obtained:

$$G_c(s)=\frac{n(s)}{k_Q(s)}=-\frac{N_0^2\cdot s}{(I-CK_D)\cdot s^2+(2K_{Q0}N_0-CK_P)\cdot s-CK_I}$$

$$\Updownarrow$$

$$G_c(s)=\frac{n(s)}{k_Q(s)}=\frac{-\dfrac{N_0^2}{I-CK_D}\cdot s}{s^2+\dfrac{2K_{Q0}N_0-CK_P}{I-CK_D}\cdot s+\left(-\dfrac{CK_I}{I-CK_D}\right)} \tag{4.41}$$

A main objective in introducing the PID (or any other) control scheme is to achieve acceptable shaft speed (rpm) regulation. In turn, this means that satisfactory propeller load disturbance rejection (attenuation) must be obtained, by proper selection of the PID control law gains K_P, K_I and possibly K_D. As explained in Appendix B, disturbance rejection of any transfer function is quantified by use of the H∞-norm. Therefore, the generic requirement has the following form for $G_c(s)$:

$$\|G_c(s)\|_\infty \le G_0 \tag{4.42}$$

In Appendix B it is argued that the above requirement can be met if specific values are assigned to the coefficients of the closed-loop transfer function $G_c(s)$. In the case of the second-order transfer function examined earlier, the coefficients of $G_c(s)$ include steady-state gain K, damping ratio ζ and natural underdamped frequency ω_n. The latter two also determine the poles $s_{1,2}$ of the closed-loop characteristic polynomial. Therefore, the problem is essentially translated to one of loop shaping.

In effect, determination of the PID controller gains is done by equating the two expressions of the coefficients of the closed-loop transfer function. Indeed:

$$\frac{K\omega_n^2\cdot s}{s^2+2\zeta\omega_n\cdot s+\omega_n^2}=G_c(s)=\frac{-\dfrac{N_0^2}{I-CK_D}\cdot s}{s^2+\dfrac{2K_{Q0}N_0-CK_P}{I-CK_D}\cdot s+\left(-\dfrac{CK_I}{I-CK_D}\right)} \tag{4.43}$$

Therefore, the following set of the three algebraic equations for the controller gains K_P, K_I and K_D are obtained (parameters K, ζ and ω_n are specified for meeting the H∞-norm requirement, and therefore treated as given):

$$\begin{aligned} -\frac{N_0^2}{I - CK_D} &= K\omega_n^2 \\ \frac{2K_{Q0}N_0 - CK_P}{I - CK_D} &= 2\zeta\omega_n \\ -\frac{CK_I}{I - CK_D} &= \omega_n^2 \end{aligned} \tag{4.44}$$

Before concluding this section some algebraic manipulation of the equations above is done. The first equation is resolved with respect to K_D, yielding:

$$K_D = \frac{1}{C} \cdot \left(I + \frac{N_0^2}{K\omega_n^2} \right) \tag{4.45}$$

The second and third equations are divided into parts, with the first one giving the following equations for K_P and K_I respectively:

$$\left. \begin{aligned} \frac{CK_P - 2K_{Q0}N_0}{N_0^2} &= \frac{2\zeta}{K\omega_n} \\ \frac{CK_I}{N_0^2} &= \frac{1}{K} \end{aligned} \right\} \Leftrightarrow \left\{ \begin{aligned} K_P &= \frac{2N_0}{C} \cdot \left(K_{Q0} + \frac{\zeta N_0}{K\omega_n} \right) \\ K_I &= \frac{N_0^2}{K \cdot C} \end{aligned} \right. \tag{4.46}$$

The above two equations can be used for the calculation of the proportional and integral gain, K_P and K_I respectively, in the case of either PI or PID control. The two cases are examined more closely and customised for a typical propulsion powerplant of a large containership in the next section; their performance is assessed comparatively and some important conclusions are made.

4.4 PI and PID H-infinity Regulation of Shaft RPM

4.4.1 Overview and Requirements

The objective of this section is to investigate the performance of the PI and PID shaft rpm regulators of the typical marine plant; the control law gains have been calculated using the three design algebraic equations derived in the previous section for meeting H∞-norm requirements with proper adaptations. Both regulators are coupled to the typical marine propulsion plant of the large containership "Shanghai Express" introduced in Chapter 3. In effect, fruitful conclusions are made leading to the choice of the PID regulator as being best suited for disturbance rejection near MCR and under severe propeller fluctuation operating conditions. The practical problems arising due to the incorporation of the D-term are examined in the next section, where a method for dealing with them is proposed.

The current section concludes with fitting both the H∞ regulators in the MoTher detailed engine simulation platform. The improved transient response is demonstrated to coincide with the one obtained in Matlab/Simulink using the localised empirical transfer functions (both full and reduced-order) presented in the previous chapter for the same plant.

A major design choice for both the H∞ regulators is to obtain a closed-loop transfer function with $\zeta = 1.0$, *i.e.* a double real pole. This choice is based upon the analysis presented previously in order to achieve the following:

- step response is not oscillatory
- relative stability is not compromised
- calculation of the H∞-norm of the closed-loop transfer function is not algebraically cumbersome.

Therefore, ζ will be assumed unity in both cases of the PI and PID H∞ rpm regulator.

4.4.2 The PI H∞ RPM Regulator

In this case: $K_D = 0$. Consequently, the generic design equations are modified as follows, if the D-term is excluded from the controller and damping ratio ζ is set to 1.0:

$$
\begin{aligned}
-\frac{N_0^2}{I} &= K\omega_n^2 \\
\frac{2K_{Q0}N_0 - CK_P}{I} &= 2\omega_n \\
-\frac{CK_I}{I} &= \omega_n^2
\end{aligned}
\tag{4.47}
$$

Note, that the first of the above equations has actually been converted to a loop-shaping constraint (instead of a design equation), as the regulator employed is a PI one.

The following H∞-norm requirement for the closed-loop transfer function $G_c(s)$ is considered:

$$\|G_c(s)\|_\infty \leq G_0$$

For design purposes, the equality part of the above relationship will be held, as:

- the controller should not be made too conservative, as this leads to increased controller gains, and

- the inequality part can, in any case, be met by satisfying the equality part for a value G_0' for $\|G_c(s)\|_\infty$, provided that $G_0' < G_0$.

Taking into account that $\zeta = 1.0$:

$$\|G_c(s)\|_\infty = |G_c(j\omega_n)| = \frac{|K|\omega_n}{2} \tag{4.48}$$

In effect, by resolving the equality of the H∞-norm requirement, the following loop-shaping constraint equation is obtained:

$$\|G_c(s)\|_\infty = G_0 \Rightarrow \frac{|K|\omega_n}{2} = G_0 \Rightarrow |K|\omega_n = 2G_0 \tag{4.49}$$

By rewriting the first design equation, the natural underdamped frequency ω_n is calculated as follows:

$$K\omega_n^2 = -\frac{N_0^2}{I} \Rightarrow |K|\omega_n \cdot \omega_n = \frac{N_0^2}{I} \Rightarrow \omega_n = \frac{N_0^2}{2G_0 I} \tag{4.50}$$

Furthermore, the value of the steady-state gain K is obtained:

$$K\omega_n^2 = -\frac{N_0^2}{I} \Rightarrow K = -\frac{N_0^2}{I\omega_n^2} \Rightarrow K = -\frac{4G_0^2}{N_0^2} \cdot I \tag{4.51}$$

Therefore, the parameters of the standard form of the closed-loop transfer function:

$$G_c(s) = \frac{K\omega_n^2 \cdot s}{s^2 + 2\zeta\omega_n \cdot s + \omega_n^2}$$

have been calculated based on the H∞-norm requirement. Note that steady-state gain is negative. However, this does not introduce any modifications in the conclusions made for the magnitude Bode plot of $G_c(s)$ using the paradigms of the previous section (for which steady-state gain is positive unity), as only $|K|$ is involved in calculations for $|G_c(j\omega)|$. On the other hand, negative steady-state gain introduces a phase shift of π (or equivalently $(-\pi)$) in the phase Bode plots; however, this does not affect the shape of the phase plot otherwise.

Having determined the parameters of $G_c(s)$ for satisfying the H∞-norm requirement, the PI regulator gains can be readily calculated as follows:

$$\begin{aligned} K_P &= \frac{2N_0}{C} \cdot \left(K_{Q0} + \frac{\zeta N_0}{K\omega_n} \right) = \frac{N_0}{C} \cdot \left(2K_{Q0} - \frac{N_0}{G_0} \right) \\ K_I &= \frac{N_0^2}{K \cdot C} = -\frac{N_0^4}{4IG_0^2 \cdot C} \end{aligned} \tag{4.52}$$

In conclusion, the closed-loop propulsion system, using a PI H∞ rpm regulator with the gains above, obtains a double real pole at $s_{1,2} = -\omega_n = -(N_0^2 / 2G_0 I)$ and satisfies the disturbance rejection requirements, as the H∞-norm of the closed-loop transfer function is set to be G_0.

4.4.3 The PID H∞ RPM Regulator

In this case all three terms are available in the PID control law. Consequently, the generic design equations can be used with the damping ratio ζ set to 1.0:

$$\left.\begin{aligned} -\frac{N_0^2}{I-CK_D} &= K\omega_n^2 \\ \frac{2K_{Q0}N_0 - CK_P}{I-CK_D} &= 2\omega_n \\ -\frac{CK_I}{I-CK_D} &= \omega_n^2 \end{aligned}\right\} \Rightarrow \left\{\begin{aligned} K_D &= \frac{1}{C}\cdot\left(I + \frac{N_0^2}{K\omega_n^2}\right) \\ K_P &= \frac{2N_0}{C}\cdot\left(K_{Q0} + \frac{N_0}{K\omega_n}\right) \\ K_I &= \frac{N_0^2}{K\cdot C} \end{aligned}\right. \tag{4.53}$$

Note that in this case the first of the above equations offers a degree of freedom to design, instead of imposing an additional loop-shaping constraint, as in the case of the PI regulator. Therefore, it is not necessary to use the value of natural underdamped frequency ω_n in order to meet the H∞-norm requirement; this is explained below.

By using the same arguments as in the case of the PI regulator, the following loop-shaping constraint equation is derived from the H∞-norm requirement, also by taking into account that $\zeta = 1.0$:

$$\left.\begin{aligned} \|G_c(s)\|_\infty &= |G_c(j\omega_n)| = \frac{|K|\omega_n}{2} \\ \|G_c(s)\|_\infty &= G_0 \end{aligned}\right\} \Rightarrow |K|\omega_n = 2G_0 \Rightarrow |K| = 2G_0/\omega_n \tag{4.54}$$

Consequently, the natural underdamped frequency is left to be chosen freely. As explained, choice is based upon relative stability, as well as additional loop-shaping considerations.

As mentioned previously, the value of the natural underdamped frequency actually determines the relative stability in the case where damping ratio is set to unity. Indeed, the double real pole of the transfer function, when $\zeta = 1.0$, is located at $s_{1,2} = -\omega_n$. Therefore, increasing ω_n is actually translated to increased relative stability, as the pole moves deeper in the left-half s-plane.

However, concerns other than relative stability should also be taken into account, such as robustness against neglected dynamics, which is examined in the following paragraph. Furthermore, proper shaping of the closed-loop transfer function $\tilde{G}_c(s)$ from propeller load disturbance k_Q to fuel index perturbation signal f_R is important, as demonstrated next. Calculation of $\tilde{G}_c(s)$ proceeds as follows:

$$f_R(s) = \left(K_P + \frac{K_I}{s} + K_D s\right)\cdot n(s) = \frac{K_D s^2 + K_P s + K_I}{s}\cdot n(s) \tag{4.55}$$

and

$$n(s) = \frac{C}{Is + 2K_{Q0}N_0} \cdot f_R(s) - \frac{N_0^2}{Is + 2K_{Q0}N_0} \cdot k_Q(s) \tag{4.56}$$

Thus:

$$f_R(s) = \frac{K_D s^2 + K_P s + K_I}{s} \cdot \left\{ \frac{C}{Is + 2K_{Q0}N_0} \cdot f_R(s) - \frac{N_0^2}{Is + 2K_{Q0}N_0} \cdot k_Q(s) \right\}$$

$$\Updownarrow$$

$$\tilde{G}_c(s) = \frac{f_R(s)}{k_Q(s)} = -\frac{N_0^2 \cdot \left(K_D s^2 + K_P s + K_I\right)}{(I - CK_D) \cdot s^2 + (2K_{Q0}N_0 - CK_P) \cdot s - CK_I} \tag{4.57}$$

The following remarks can be made concerning $\tilde{G}_c(s)$ calculated above. (i) The characteristic polynomial $p_c(s)$ of $\tilde{G}_c(s)$ is the same as that of $G_c(s)$ and, therefore, obtains identical poles. (ii) In the PI regulator case $(K_D = 0)$ the transfer function $\tilde{G}_c(s)$ is always minimum-phase, provided that $(K_P \cdot K_I) > 0$, which is normally the case. (iii) In the PID regulator case the transfer function $\tilde{G}_c(s)$ may be non-minimum-phase, *i.e.* obtain at least one zero in the right-half s-plane.

Remark (ii) is derived readily by noticing that when $K_D = 0$, the numerator polynomial $p_n(s)$ becomes:

$$p_n(s) = K_P s + K_I \Rightarrow z_0 = -\frac{K_I}{K_P} \tag{4.58}$$

Therefore, $(K_P \cdot K_I) > 0 \Rightarrow z_0 < 0$, implying that the closed-loop transfer function $\tilde{G}_c(s)$ is minimum-phase.

Remark (iii) refers to the PID regulator case; the introduction of the D-term increases the order and the number of roots of the numerator polynomial, $p_n(s)$, to two. This makes possible that at least one of the roots lies in the right-half s-plane, causing $\tilde{G}_c(s)$ to become non-minimum-phase. However, as said in [33], a necessary, but not sufficient, condition for a polynomial of arbitrary order with real coefficients not to obtain roots in the right-half s-plane is that all coefficients of the polynomial have the same sign and do not vanish. In the case of second-order polynomials, such as $p_n(s)$, it can be easily seen from the analytical expressions for the roots that the above condition is also sufficient.

In normal cases, proportional and integral gains, K_P and K_I respectively, are negative (note that in this work the rpm perturbation n is considered to be the feedback signal, instead of the tracking error e of shaft rpm, and it holds that $n = -e$). Therefore, a necessary and sufficient condition for $\tilde{G}_c(s)$ to be minimum-phase is that $K_D < 0$ as well. From the H∞-norm requirement, and assuming without any loss of generality that $K < 0$, we obtain that:

$$K = -\frac{2G_0}{\omega_n} \tag{4.59}$$

By substituting the above value of K in the first design equation, in order to resolve with respect to ω_n :

$$-\frac{2G_0}{\omega_n} \cdot \omega_n^2 = -\frac{N_0^2}{I - CK_D} \Rightarrow \omega_n = \frac{N_0^2}{2G_0} \cdot \frac{1}{(I - CK_D)} \tag{4.60}$$

By defining $\omega_{n,\mathrm{PI}}$ the value of ω_n for $K_D = 0$ (*i.e.* $\omega_{n,\mathrm{PI}} = N_0^2 / (2G_0 I)$), the following constraint inequality is obtained for natural underdamped frequency:

$$K_D < 0 \Leftrightarrow \omega_n < \omega_{n,\mathrm{PI}} \tag{4.61}$$

Consequently, using the argumentation above:

$$\tilde{G}_c(s) \text{ is minimum-phase} \Leftrightarrow \omega_n < \omega_{n,\mathrm{PI}} \tag{4.62}$$

Therefore, when the PID control law is employed instead of the PI one, relative stability has to be compromised in order to guarantee that transfer function $\tilde{G}_c(s)$ remains minimum-phase. Another major concern during PID controller design is robustness against neglected dynamics, which is discussed next.

4.4.4 Robustness Against Neglected Dynamics

It is generally known that (stable) poles having a dominant effect on transient response (dominant poles) are the ones which are located closer to the imaginary axis (have smaller real parts in the absolute sense) [10, 33]. Since most practical systems obtain open-loop transfer functions with order higher than two, it is most useful to use only a few (preferably just one) dominant poles for the controller design. In the literature [33], it is mentioned that poles with real part five to ten times smaller than any other can be considered as dominant.

In the case of marine plants, dynamical terms due to turbocharging, fuel injection delay, and the presence of an actuator have been neglected during the H∞ procedure presented previously. Indeed, the reduced-order open-loop transfer function of order one has been considered, which gives a closed-loop transfer function with order two, when a PI or a PID controller is coupled to the plant. Thus, it is imperative also to consider the properties of robust stability and performance for the closed-loop system, when the neglected dynamics are taken into account.

In [33] it is referred that the dominant pole approach to control design is robust against the neglected, insignificant dynamics (especially the open-loop poles), if adequate "safety" distance is provided between the insignificant poles and the dominant ones. Specifically, the left-half s-plane is sectionalised into regions in which dominant and less dominant poles are lying or may lie after pole-placement due to controller coupling. These two regions are bounded by straight lines parallel to the imaginary axis; the distance between the boundary straight lines is actually the safety distance between dominant and insignificant poles.

To illustrate the effect of neglected dynamics, in the case of typical marine propulsion powerplants, an analysis for the effect of the turbocharging term $(1+\tau_{TC}s)$, neglected during controller design, is given. First, the transfer function is extended so that the neglected dynamical term is included.

The open-loop transfer function is:

$$n(s)=\frac{C}{(\tau_{TC}s+1)\cdot(Is+2K_{Q0}N_0)}\cdot f_R(s)-\frac{N_0^2}{Is+2K_{Q0}N_0}\cdot k_Q(s) \tag{4.63}$$

or more compactly:

$$n(s)=\breve{G}_1(s)\cdot f_R(s)+G_2(s)\cdot k_Q(s)$$

By introducing the generic PID control law:

$$f_R(s)=K_c(s)\cdot n(s)=\left(K_P+\frac{K_I}{s}+K_D s\right)\cdot n(s) \tag{4.64}$$

the following modified closed-loop transfer function $G_c'(s)$ from the propeller disturbance signal k_Q to shaft rpm fluctuation n is obtained:

$$G_c'(s)=\frac{n(s)}{k_Q(s)}=\frac{G_2(s)}{1-\breve{G}_1(s)\cdot K_c(s)}$$

$$\Updownarrow$$

$$G_c'(s)=-\frac{N_0^2\cdot s\cdot(1+\tau_{TC}\cdot s)}{\tau_{TC}I\cdot s^3+(I-CK_D+\tau_{TC}\cdot 2K_{Q0}N_0)\cdot s^2+(2K_{Q0}N_0-CK_P)\cdot s-CK_I} \tag{4.65}$$

The following remarks can be made concerning transfer function $G_c'(s)$ in correlation with $G_c(s)$:

- the real and negative closed-loop zero $(-1/\tau_{TC})$ is introduced
- the order of the closed-loop transfer function is increased to three, as expected.

The most important effect on the transient performance, however, originates from the modification of the closed-loop characteristic polynomial. For further analysis, the following error polynomial $\varepsilon_c(s)$ is defined:

$$\varepsilon_c(s)=p_c'(s)-p_c(s)=\tau_{TC}s^2\cdot(Is+2K_{Q0}N_0) \tag{4.66}$$

where $p_c'(s)$ is the characteristic polynomial of transfer function $G_c'(s)$:

$$\begin{aligned}p_c'(s)&=a_3's^3+a_2's^2+a_1's+a_0'\\&=\tau_{TC}I\cdot s^3+(I-CK_D+\tau_{TC}\cdot 2K_{Q0}N_0)\cdot s^2+(2K_{Q0}N_0-CK_P)\cdot s-CK_I\end{aligned} \tag{4.67}$$

$\varepsilon_c(s)$ give a double root at $s=0$ and a single one at $s=-2K_{Q0}N_0/I=-\tau_{prop}^{-1}$, *i.e.* at the dominant open-loop pole of the marine plant transfer function. Therefore:

- the introduction of the fast dynamical term due to turbocharging does not affect steady-state performance, as $\varepsilon_c(0+j0)=0$
- as expected, when controller gains are set to zero $(K_c(s)\equiv 0)$ both $G_c'(s)$ and $G_c(s)$ coincide with open-loop transfer function $G_2(s)$ from propeller disturbance k_Q to shaft rpm fluctuation n as $\varepsilon_c(-\tau_{prop}^{-1})=0$.

In general, care should be taken during selection of controller gains, so that the value of the error polynomial $\varepsilon_c(s)$ at the poles assigned to the characteristic polynomial $p_c(s)$ of the reduced-order, closed-loop transfer function is adequately small. If this is the case, then it means that two of the roots of $p_c'(s)$ (third-order polynomial) are in a close vicinity to the roots of $p_c(s)$ (second-order polynomial), due to continuity of polynomial functions.

According to the H∞ design procedure deployed previously for marine plants, it has been selected to obtain a closed-loop characteristic polynomial with a double real root at $s_{1,2}=-\omega_n$ This means that:

$$p_c(s)=(I-CK_D)\cdot(s^2+2\omega_n s+\omega_n^2) \tag{4.68}$$

Also, the following general relationship holds:

$$p_c'(s)=(a_3's+(a_2'-2a_3'\omega_n))\cdot(s^2+2\omega_n s+\omega_n^2)+\upsilon(s) \tag{4.69}$$

where $\upsilon(s)$ is the residual first-order polynomial of the division. Then, provided that either:

- $\omega_n \ll$, or
- $\omega_n\cdot\tau_{prop}\approx 1$,

$\varepsilon_c(s=-\omega_n)$ obtains an adequately small value, as explained earlier; thus it can be assumed that $s_{1,2}=-\omega_n$ is also a double root of $p_c'(s)$. This, in turn, implies that $\upsilon(s)$ can be omitted from the division equation above. Therefore, the third root of $p_c'(s)$ is calculated as the root of the quotient polynomial:

$$a_3's+(a_2'-2a_3'\omega_n)=0\Rightarrow s_3=-\frac{a_2'-2a_3'\omega_n}{a_3'} \tag{4.70}$$

After some algebraic manipulation, it is obtained that:

$$s_3=-\frac{I-CK_D-2\tau_{TC}\cdot(I\omega_n-K_{Q0}N_0)}{\tau_{TC}I}=-\frac{1}{\tau_{TC}}\cdot\left[1-\frac{\tau_{TC}}{\tau_{prop}}\cdot\left(2\omega_n\tau_{prop}-1\right)-\frac{C}{I}\cdot K_D\right] \tag{4.71}$$

Relative stability considerations make the design choice of $\omega_n \cdot \tau_{prop} \approx 1$ more preferable than selecting a small value for ω_n. Therefore, $\varepsilon_c(s=-\omega_n)$ can be made adequately small by placing ω_n as close to $1/\tau_{prop}$ as required. Note that, in the H∞ PID regulator case. ω_n can be adjusted in order to fulfil the above requirement $\omega_n \cdot \tau_{prop} \approx 1$, whilst in the H∞ PI regulator case this is not possible, as the value of ω_n is fixed in order to meet the H∞-norm requirement. Furthermore, by assuming $\omega_n \cdot \tau_{prop} \approx 1$ the expression for s_3 becomes:

$$s_3 = -\frac{1}{\tau_{TC}} \cdot \left(1 - \frac{\tau_{TC}}{\tau_{prop}} - \frac{C}{I} \cdot K_D \right) \tag{4.72}$$

In effect, the following conclusions can be made:

H∞ PI regulator: As $K_D = 0$:

$$s_{3,\mathrm{PI}} = -\frac{1}{\tau_{TC}} \cdot \left(1 - \frac{\tau_{TC}}{\tau_{prop}} \right) > -\frac{1}{\tau_{TC}} \tag{4.73}$$

meaning that closed-loop pole s_3 is inescapably closer to the dominant pole region, compared with the open-loop, insignificant pole $-1/\tau_{TC}$, depicting turbocharging and combustion dynamics of the marine plant. Therefore, even if the value of ω_n provides $\omega_n \cdot \tau_{prop} \approx 1$, performance robustness of the closed-loop system is compromised.

H∞ PID regulator: As $K_D < 0$:

$$s_3 = -\frac{1}{\tau_{TC}} \cdot \left(1 - \frac{\tau_{TC}}{\tau_{prop}} - \frac{C}{I} \cdot K_D \right) < s_{3,\mathrm{PI}} \tag{4.74}$$

Therefore, performance robustness of the closed-loop system is compromised less than in the H∞ PI case. Furthermore, as in the typical case it can be assumed that:

$$\frac{\tau_{TC}}{\tau_{prop}} < \frac{C}{I} \cdot |K_D| \tag{4.75}$$

it can be concluded that:

$$s_3 < -\frac{1}{\tau_{TC}} \tag{4.76}$$

i.e. the third pole of $G_c'(s)$ is faster than the open-loop pole introduced by the turbocharging term, and in effect belongs to the insignificant region of the left-half complex s-plane. Therefore, the approximation of $G_c'(s)$ using $G_c(s)$, especially for $s \to 0$, is acceptable, and the PID H∞ controller designed using the reduced-order transfer function of the marine plant is robust against the neglected dynamics of the turbocharging term.

It has been shown that the freedom to choose the value of natural underdamped frequency ω_n may prove rather necessary, as if it is placed far from the open-loop dominant pole, determined by τ_{prop}, closed-loop error polynomial $\varepsilon_c(s)$ may give a large value at $s=-\omega_n$, leading to large deviations in closed-loop performance. Additionally, the D-term, in spite of any practical problems envisaged in implementation, increases the safety distance of closed-loop pole s_3 from the dominant region, leading to enhanced robustness against the turbocharger dynamical term. This is illustrated in the typical numerical case investigation presented below, where the PI and PID H∞ controllers are fit to the propulsion plant of a large containership.

Moreover, as seen above, the theoretical analysis of robustness against neglected dynamics is a rather cumbersome process and, as explained in [33], it is not an exact science. In particular, if all neglected dynamical terms (not just one of them as in the procedure above) are to be taken into account, trial-and-error techniques have to be used employing simulation, because the order of the closed-loop polynomial is growing larger. This is also demonstrated in the typical example below.

4.4.5 Numerical Investigation of a Typical Case

The PID rpm regulator tuning methodology described above has been applied to the rpm regulation of the propulsion powerplant of containership "Shanghai Express", which is equipped with an MAN-B&W 9K90MC-VI marine engine. The performance details, as well as open-loop transfer function identification of this specific plant, were given in Section 3.3.5. Therefore, they are not repeated here.

The step following the open-loop modelling and identification is to define the numerical value for $\|G_c(s)\|_\infty$ that guarantees satisfactory closed-loop performance, in terms of disturbance rejection. To meet this objective, the maximum absolute deviation, δk_Q, of propeller torque coefficient, K_Q, from the nominal design value for the propeller, $K_{Q0}=0.4753\,\text{kN m/rpm}^2$ has to be determined. Propeller torque coefficient, however, is not directly measurable onboard. On the other hand, shaft load torque is usually recorded in conjunction with shaft rpm. Thus, by taking into account the speed governor settings, fluctuation magnitude δk_Q can be deduced. Specifically, if the maximum absolute propeller torque fluctuation $\delta Q_L \triangleq \max|Q_L-Q_{L0}|=\max\left|Q_L-K_{Q0}\cdot N_0^2\right|$ is known, then by using:

$$Q_L-Q_{L0}=2K_{Q0}N_0\cdot n+N_0^2\cdot k_Q$$

the maximum absolute deviation, δk_Q, can be estimated as follows:

$$\delta k_Q = \frac{\delta Q_L}{N_0^2} \tag{4.77}$$

Note that the above estimation of δk_Q is rather conservative, as it is assumed that, for propeller torque fluctuation, only torque coefficient deviation k_Q is to be held responsible; the contribution of shaft rpm n in propeller torque fluctuation is omitted.

A campaign of shipboard measurements has been carried out by NTUA-LME, onboard "Shanghai Express" [4,27]. The data series gathered indicated that propeller torque fluctuation with magnitude 300 kN m may occur under certain operating conditions. For the MAN-B&W 9K90MC-VI marine engine the rpm limit is 96.0 rpm. Therefore, overspeed occurs when the magnitude of rpm fluctuation exceeds 1.3 rpm, if the rpm setpoint is $N_0 = 94.7$ rpm. Therefore, δk_Q is:

$$\delta k_Q = \frac{300 \text{ kN m}}{(94.7 \text{ rpm})^2} = 0.0335 \text{ kN m/rpm}^2$$

In effect, an upper bound value for the magnitude Bode plot is:

$$|G_c(j\omega)| \le \frac{1.3 \text{ rpm}}{0.0335 \text{ kN m/rpm}^2} = 38.806 \text{ rpm}^3\text{/kN m or } 20\log(38.806) = 31.778 \text{ dB}$$

Being a little more conservative, the specification for $\|G_c(s)\|_\infty = G_0$ is set to 26.0 dB, which corresponds to a maximum rpm deviation of 1 rpm (approximately 1%MCR) for propeller torque coefficient deviation of 0.05 kN m/rpm^2 (approximately 10%K_{Q0} or 10%Q_{MCR} in terms of propeller torque fluctuation).

The tuning procedure has been applied for both cases of an H∞ PI and PID controller (referred to hereafter as Hinf PI and Hinf PID respectively). In addition, for reference purposes the response of the plant with the standard PI (std PI) controller, used for rpm regulation at the actual shipboard installation, is presented. The gains of the three regulators are given in the Table 4.1. Note the negative signs in front of the PID constants, as, most commonly, the speed error (setpoint minus actual rpm) is the governor's driving signal, instead of rpm fluctuation, as defined in the text.

Table 4.1 PID controller gains for the "Shanghai Express" powerplant

	$(-K_P)$ (%index/rpm)	$(-K_I)$ (%index/rpm/s)	$(-K_D)$ (%index · s/rpm)
std PI	5.00	1.00	0.00
Hinf PI	13.19	22.79	0.00
Hinf PID	13.19	11.67	2.53

Comparison between the three PID regulators can be done using the Bode plots of the closed-loop transfer function derived in each case shown in Figure 4.8.

As already mentioned, the relatively most stable system is the one using the Hinf PI. The Hinf PID system is more stable than the std PI but less than the Hinf PI one.

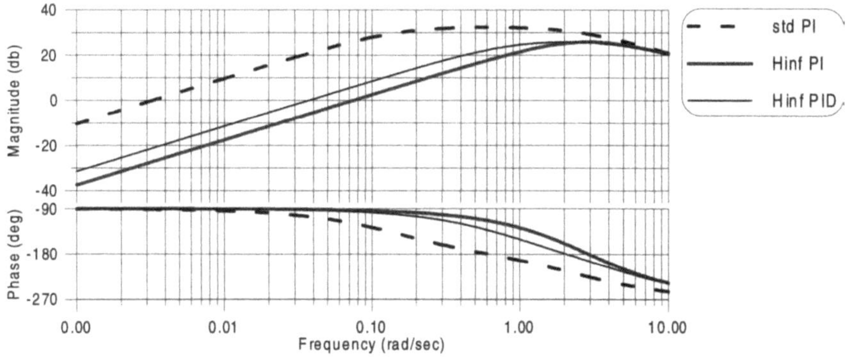

Figure 4.8 Bode plots of the reduced-order plant with std PI, Hinf PI and PID rpm regulators

In the case of std PI the closed-loop transfer function obtains damping ratio $\zeta = 2.27$ and natural underdamped frequency $\omega_n = 0.61$ rad/s. However, disturbance rejection with std PI is rather poor, as the peak of the magnitude Bode plot, occurring at $\omega = \omega_n$, exhibits a value of approximately 32.5 dB, *i.e.* 6.5 dB deviation from the specification. Therefore, safe operation near-MCR under significant load demand fluctuation cannot be ensured using the std PI.

In the case of Hinf PI, loop-shaping provides the following values for damping ratio, natural underdamped frequency and steady-state gain: $\zeta = 1.0$, $\omega_n = 2.93$ rad/s and $K = -2.33$. Note that, except for the smaller (and acceptable) value achieved for $\|G(s)\|_\infty$, the peak of the magnitude plot has also been shifted to a larger frequency, *i.e.* the system is relatively more stable. This is reflected by the step response as well, and is preferable because the load disturbance psd distribution is expected to be a decreasing function of frequency. However, as already said, an increase of relative stability is not without a price. Indeed, enhanced relative stability leads to a closed-loop system with significantly reduced robustness against neglected dynamics.

In the case of the Hinf PID regulator, incorporation of the D-term in the control law gives an additional degree of freedom, as there is one more design parameter available. In the test case examined, and under the considerations presented previously, the D-term was used for providing ω_n approximately half the value of the Hinf PI case, *i.e.* $\omega_n = 1.50$ rad/s. Note that P-constant remains unaltered and I-constant is halved compared to the Hinf PI case. The latter is a direct consequence of halving ω_n.

For validation purposes, the sinusoidal steady-state response of the three transfer functions is shown in Figure 4.9 for ω_n values of 0.61, 1.50 and 2.93 rad/s, *i.e.* at the natural underdamped frequencies of the three closed-loop transfer functions. The compliance of both Hinf regulators to specification is demonstrated in Figure 4.9.

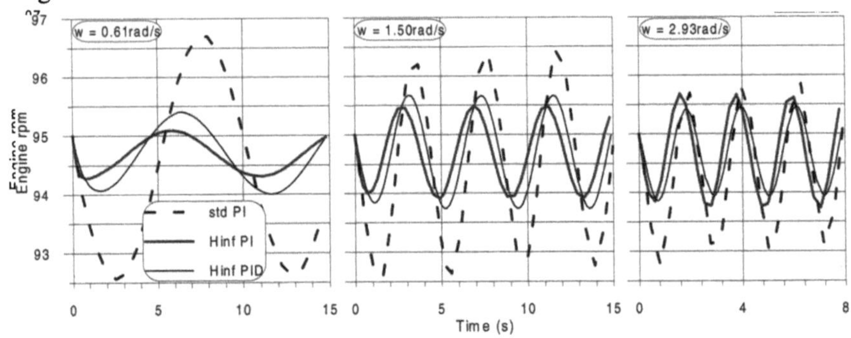

Figure 4.9 Sinusoidal rpm response of the three systems at natural underdamped frequencies

Finally, the step response to a 0.05 kN m/rpm^2 step in k_Q of the reduced-order system (no neglected dynamics included) with the Hinf regulators is shown in Figure 4.10. As can be seen, the performance is rather satisfactory and comparable for both of them. Maximum overspeed for both regulators is practically the same. On the other hand, the more stable pole introduced by the Hinf PI regulator gives a significantly faster response. In any case, either one of these regulators can guarantee performance for the reduced-order plant, whilst the std PI cannot.

Using the speed governor element available in the MoTher engine simulation tool, validation of the three regulators has been carried out. Figure 4.11 shows the step and sinusoidal steady-state response obtained using disturbances like the ones applied to the reduced-order transfer function. The frequency of the sinusoid applied was 0.61rad/s. Note the high-frequency process noise propagated in the fuel index signal, when the Hinf PID regulator is used, which is due to direct differentiation of the rpm signal.

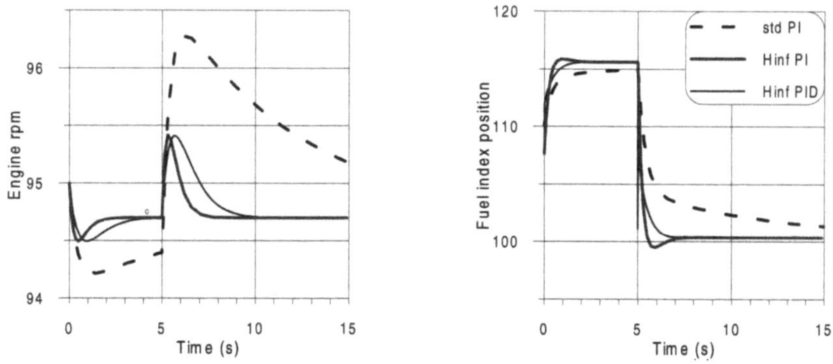

Figure 4.10 Response of the reduced-order plant with std PI, Hinf PI and Hinf PID regulators

However, after two higher-order, fast dynamical terms are introduced to the system, only the Hinf PID maintains acceptable performance levels. Specifically, a turbocharging term with $\tau_{TC} = 0.25\ \mathrm{s}$ and an actuator term $\tau_{act} = 0.10\ \mathrm{s}$ are included in the full-order transfer function of the marine plant. Then, as seen in Figure 4.12, although the observed increase of overspeed with the Hinf PID is not negligible, the closed-loop system step response is still maintained approximately to 1 rpm for a step in k_Q of 0.05 kN m/rpm^2. Furthermore, the response does not exhibit any oscillations. On the other hand, the performance of the Hinf PI after enrichment of dynamics becomes unacceptable. As seen in Figure 4.12, the engine rpm exceeds the ovespeed limit of 96.0 rpm. Additionally, significant oscillation is introduced to the closed-loop system response, giving jiggling in the fuel index signal as well. This dramatic deterioration of the Hinf PI regulator performance can be explained by the fact that the Hinf PI-compensated, reduced-order transfer function has a double real pole located deeper in the left-half complex *s*-plane. This reduces the "safety distance" between the dominant (double) pole and the poles introduced by neglected fast dynamics, such as actuator, combustion, *etc.*

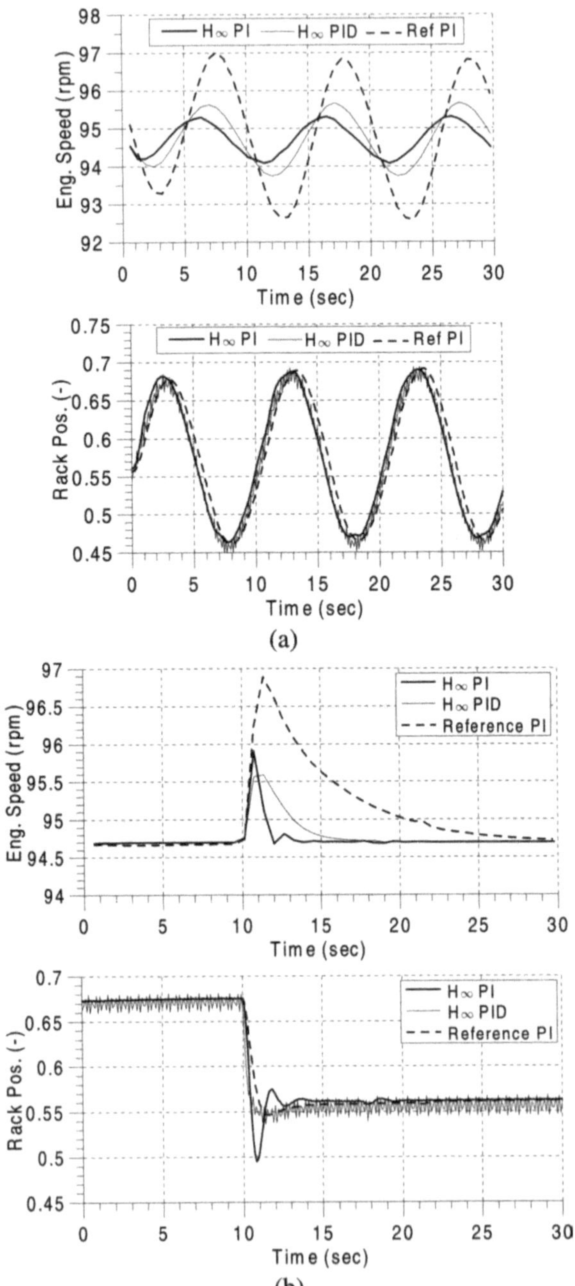

Figure 4.11 Sine (a) and step (b) response of MoTher using the three regulators – plots include rpm response, as well as fuel index response

In effect, there exists a stability and robustness against neglected dynamics trade-off that is encountered in a variety of control loops. In the case of marine propulsion the introduction of a D-term control action component allows one to compromise between closed-loop stability and robustness, as free selection of natural underdamped frequency is made possible.

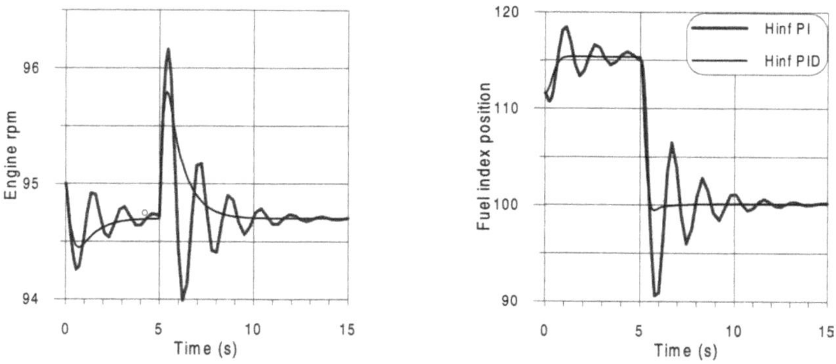

Figure 4.12 Step response of the full-order plant with std PI, Hinf PI and Hinf PID regulators

Therefore, although the reduced-order performance of both Hinf regulators is acceptable, the lack of robustness against neglected dynamics leads to the exclusion of the Hinf PI rpm regulator from realisation. However, it is well known that utilisation of the D-term, such as the one employed in the Hinf PID controller, suffers in actual installation from high-frequency noise introduced to both the manipulated variable and control input due to differentiation. Therefore, although the theoretical analysis has proved that Hinf PID is superior to Hinf PI as far as robustness is concerned, the Hinf PID cannot be used if a solution for the isolation of high-frequency noise is not found.

4.5 D-term Implementation Using Shaft Torque Feedback

4.5.1 Real-time Differentiation and Linear Filters

The degree of freedom offered by the inclusion of the D-term in the control law has proven crucial for achieving an rpm regulator of the PID type that can offer disturbance rejection and ensure robustness against high-order fast dynamics, which have been neglected in the design procedure. The D-term incorporation in the control law, however, introduces a new type of practical problem due to the amplification of high-frequency noise. In particular, in the case of marine propulsion powerplants, in addition to the measurement error and instrumentation noise, a component of process noise due to the discrete cylinder firings and the finite number of propeller blades is introduced in the shaft rpm signal.

It is now argued that a linear low-pass filter, placed in the feedback path of the shaft rpm signal cannot eliminate the high-frequency noise in the rpm derivative to a desirable level. Indeed, the frequency (magnitude) characteristic of a linear filter can be approximated by the following simplified expression [36–38]:

$$H_{LPF}(\omega) = \begin{cases} 1 & ,\omega \le 2\pi N_0/60 \\ \varepsilon_{\mathrm{LPF}} & ,\omega > 2\pi N_0/60 \end{cases} \tag{4.78}$$

Note that this expression is just an approximation, as linear, proper transfer functions, *i.e.* implemented as a ratio of two polynomials with the degree of the numerator smaller than or equal to that of the denominator polynomial:

- are continuous, and therefore a transition band is placed between the passband and stopband; therefore the discontinuity of $H_{LPF}(\omega)$ at $\omega = 2\pi N_0/60$ is not realisable with a linear filter;
- have, in general, nonzero derivatives with respect to frequency over both the stopband and passband; therefore the attenuation is demonstrating fluctuation over both bands instead of being a constant (*e.g.* 1.0).

However, the most important factor is that, by using a linear transfer function for the filter implementation, the gain in the stopband $\varepsilon_{\mathrm{LPF}}$ cannot be equal to zero, except from several (enumerated) values of frequency ω that are determined by the zeros of $H_{LPF}(\omega)$. However, $\varepsilon_{\mathrm{LPF}} \ll 1$, so that rejection of the high-frequency components meets some specification. Also, note that care is usually taken during design, so that:

$$\lim_{\omega \to +\infty} H_{LPF}(\omega) = 0 \tag{4.79}$$

In any case, for the purposes of this text, the approximation given above will be considered in order to investigate the effect of the filter. The filtered rpm signal $\rlap{-}N$ will obtain in conclusion, magnitude given by the following as a function of ω:

$$|\rlap{-}N(\omega)| = H_{LPF}(\omega) \cdot |N(\omega)| = \begin{cases} |N(\omega)| & ,\omega \le 2\pi N_0/60 \\ \varepsilon_{\mathrm{LPF}} \cdot |N(\omega)| & ,\omega > 2\pi N_0/60 \end{cases} \tag{4.80}$$

Therefore, frequency components in $N(\omega)$ above $(2\pi N_0/60)$ will be filtered out by use of the filter; the attenuation imposed on the stopband is $\varepsilon_{\mathrm{LPF}}^{-1} \gg 1$. These frequency components are undesirable, as they represent either process or measurement noise. However, neither signal $\rlap{-}N$ nor N can be used for generating the rpm derivative $\dot{N}$. Indeed:

$$\dot{N}(s) = s \cdot N(s) \Rightarrow \dot{N}(j\omega) = j\omega \cdot N(j\omega) \Rightarrow |\dot{N}(\omega)| = \omega \cdot |N(\omega)| \tag{4.81}$$

If the filtered signal is employed instead, for $\omega > 2\pi N_0/60$, we obtain that:

$$\left|\dot{N}(\omega)\right| = \omega \cdot \left|\not{N}(\omega)\right| = \omega \cdot \varepsilon_{\mathrm{LPF}} \cdot \left|N(\omega)\right| \tag{4.82}$$

Evidently, the effect of high-frequency component filtering is now moderated due to the introduction of the ω-factor by the operation of differentiation. Thus direct differentiation cannot be safely applied for the generation of the D-term of the controller, even on a version of the rpm signal filtered by a high-order linear filter. The derivative signal will, in any case, carry a significant portion of process and measurement noise.

On the other hand, it has been demonstrated that the use of low-pass filters reduces significantly the noise present in the feedback signals from the plant. Therefore, if a combination of feedback plant signals can be found that allows direct estimation of shaft rpm derivative, without differentiation, then the D-term can be implemented practically in a reliable manner, without the interference of noise.

4.5.2 RPM Derivative Estimation from Fuel Index and Shaft Torque

In Section 3.2 it is argued that the equation of motion of the engine–propeller shaft in a typical marine propulsion plant is:

$$\dot{N}(t) \equiv \dot{n}(t) = \frac{Q_E(t) - Q_P(t)}{I(t)} = \frac{Q_E(t) - Q_P(t)}{I_0 + \Delta I(t)} \tag{4.83}$$

The variables in the above differential equation are cycle-mean and not instantaneous. In the same sense, the rpm time derivative indicates the variation of rpm from cycle to cycle and not the in-cycle rpm variations due *e.g.* to cylinder firings. As already pointed out, in-cycle variations cannot be dealt with by the engine speed governor and constitute interference to its normal operation; therefore, there is a need to filter them out.

From the shaft's equation of motion, it is obvious that the cycle-mean rpm derivative can be estimated if the values of cycle-mean engine and propeller torque, as well as shafting system inertia are provided as plant feedback information. In Section 3.3.2, it has been argued that engine brake torque can be assumed proportional to dimensionless fuel index value (minus friction torque), especially if rpm can be maintained approximately constant around a specified steady-state value. *I.e.* for the rpm regulation problem:

$$Q_E = C \cdot F_R - Q_{fr} \tag{4.84}$$

Also, shafting inertia uncertainty ΔI can be neglected for the purpose of shaft rpm derivative estimation, because, as demonstrated later on, the expression of rpm rate of change obtained does not include the propeller inertia. Therefore, inertia I will be treated as a known constant and it is not required to be provided by feedback from the plant.

In effect, the major problem in using the shaft's equation of motion for shaft rpm derivative estimation lies in the feedback of cycle-mean propeller torque demand. Indeed, direct measurement of this variable is considered rather difficult and infeasible in practical applications. By using theoretical hydrodynamics arguments, propeller load torque can be determined if the water pressure and flow fields (spatial distributions) are known as functions of time. Such instrumentation is available only in a laboratory environment and cannot be considered as a possibility for shipboard installation, at least in the foreseeable future. Furthermore, the computational complexity of the calculations involved does not support such a real-time implementation of propeller torque demand determination.

In the framework presented above, a propeller torque demand estimation method, using the shaft torque signal instead, is proposed hereafter. This measurement signal is usually available in modern marine plants. Indeed, a torque meter is attached to the engine–propeller shaft in order to assess the power and torque performance of the main engine during the ship delivery sea trials. The principle of torque measurement is based on monitoring the shaft's end-to-end angular deformation when it is under strain. The devices most commonly employed for this purpose are strain gauges or, more rarely though, optical (laser) encoders. In any case, the instrumentation remains shipboard even after sea trials are over, because it is a comparatively low-cost piece of equipment and proves rather useful whenever engine performance assessment is required. Torque sensors provide the shaft torque signal in real-time either as a digital output (4, 8 or more seldom 16, or 32-digit binary number) or as an analogue signal (most often current-loop of 4–20 mA). Therefore, shaft torque:

$$Q_{shaft} = K_{shaft} \cdot \varphi \tag{4.85}$$

is assumed to be available to the engine speed governor.

The dynamical analysis of the shafting system, using the two-mass plus spring model of Section 3.2, has proven that shaft torque appears only in the second canonical equation, as follows:

$$I_E \cdot \dot{n}_E - I_P \cdot \dot{n}_P = Q_E + Q_P - 2 \cdot Q_{shaft} \tag{4.86}$$

Note that, for the dynamical analysis purposes, zero initial conditions have been assumed; this assumption will be maintained here as well, as the value of the left-hand expression is investigated. By taking into account that:

$$I_E \cdot \dot{n}_E(t) - I_P \cdot \dot{n}_P(t) = \mathfrak{L}^{-1}\{I_E \cdot sn_E(s) - I_P \cdot sn_P(s)\} \tag{4.87}$$

and employing the expressions for engine-side and propeller-side rpm variables, $n_E(s)$ and $n_P(s)$ respectively, that have been obtained in Section 3.2.1, and which are repeated here for compactness:

$$\begin{aligned} n_E(s) &= \frac{I_P s^2 + K_{shaft}}{s \cdot (I_E I_P \cdot s^2 + I \cdot K_{shaft})} \cdot Q_E(s) - \frac{K_{shaft}}{s \cdot (I_E I_P \cdot s^2 + I \cdot K_{shaft})} \cdot Q_P(s) \\ n_P(s) &= \frac{K_{shaft}}{s \cdot (I_E I_P \cdot s^2 + I \cdot K_{shaft})} \cdot Q_E(s) - \frac{I_E s^2 + K_{shaft}}{s \cdot (I_E I_P \cdot s^2 + I \cdot K_{shaft})} \cdot Q_P(s) \end{aligned} \tag{4.88}$$

finally we obtain that:

$$I_E \cdot \dot{n}_E(t) - I_P \cdot \dot{n}_P(t) =$$

$$\mathfrak{L}^{-1}\left\{\frac{I_E I_P \cdot s^2}{I_E I_P \cdot s^2 + K_{shaft}(I_E + I_P)} \cdot (Q_E + Q_P) + \frac{K_{shaft}(I_E - I_P)}{I_E I_P \cdot s^2 + K_{shaft}(I_E + I_P)} \cdot (Q_E - Q_P)\right\} \tag{4.89}$$

where evidently $Q_E = Q_E(s)$ and $Q_P = Q_P(s)$. Although the above expression is rather complicated, it is greatly simplified if only the low-frequency part of the spectrum is held. This is justified by the nature and assumptions of engine and propeller torque spectra that have been presented earlier. Therefore:

$$I_E \cdot \dot{n}_E(t) - I_P \cdot \dot{n}_P(t) \approx \frac{I_E - I_P}{I_E + I_P} \cdot \left(Q_E(t) - Q_P(t)\right) \text{ for } s = j\omega \to 0^+ \tag{4.90}$$

By substituting the above in the canonical equation of the shafting system, the following relationship is obtained, which is valid for $s = j\omega \to 0^+$, allowing thus the cycle-mean (*i.e.* low-frequency variation) value estimation of propeller torque demand based on engine torque delivery (fuel index) and shaft torque feedback:

$$\begin{gathered} \frac{I_E - I_P}{I_E + I_P} \cdot (Q_E - Q_P) = Q_E + Q_P - 2 \cdot Q_{shaft} \\ \Downarrow \\ Q_P = \frac{I \cdot Q_{shaft} - I_P \cdot Q_E}{I_E} \end{gathered} \tag{4.91}$$

In conclusion, the above cycle-mean propeller torque demand estimate provides a manner to assess in real-time the cycle-mean shaft rpm rate of change, by employing the shaft torque and fuel index feedback signals. *I.e.* by substituting the above estimate for propeller torque demand in the shaft's equation of motion, we obtain that:

$$\dot{N}(t) = \frac{C \cdot F_R(t) - Q_{fr} - Q_{shaft}(t)}{I_E} \tag{4.92}$$

Note that the propeller inertia $I_P(t) = 1.15 \cdot I_{prop} + \Delta I(t)$ is not included in the above expression, justifying, thus, the assumption to neglect inertial uncertainty ΔI, and consider shafting inertia I a constant. The independence from I_P is due to the fact that the estimate deduced for $\dot{N}$ corresponds to the intuitive fact that engine and propeller accelerations coincide with the acceleration experienced by the shaft's centre of mass as far as slow (cycle-to-cycle) rpm changes are concerned. Indeed, it is worth remembering here that, according to the law of motion at the engine side:

$$\dot{n}_E(t) = \frac{Q_E(t) - Q_{shaft}(t)}{I_E} = \frac{C \cdot F_R(t) - Q_{fr} - Q_{shaft}(t)}{I_E} \tag{4.93}$$

The same result could have been obtained using the quasi-steady approximation. Indeed, in steady-state:

$$\dot{n}_E(t) = \dot{n}_P(t) = 0 \Rightarrow I_E \cdot \dot{n}_E - I_P \cdot \dot{n}_P = 0 \tag{4.94}$$

Therefore:

$$Q_E + Q_P - 2 \cdot Q_{shaft} = 0 \Rightarrow Q_P = 2 \cdot Q_{shaft} - Q_E \tag{4.95}$$

By extension of the above expression in slow dynamical situations (quasi-steady assumption), we obtain that:

$$\dot{N}(t) = \frac{2 \cdot \left(Q_E(t) - Q_{shaft}(t)\right)}{I} = \frac{C \cdot F_R(t) - Q_{fr} - Q_{shaft}(t)}{I/2} \tag{4.96}$$

However, it has been argued in 3.2.1 that for marine plants: $I_E \approx I_P \approx I/2$. By substituting in the above $I/2$ with I_E the two (2) expressions for $\dot{N}$ become effectively identical.

In conclusion, a method for indirect estimation of $\dot{N}$ (*i.e.* without the need of differentiating the rpm signal) has been obtained. In Section 4.5.3 the modified gains of the H∞ PID engine governor are calculated, for the case where shaft torque feedback is available. Then the theoretical results are evaluated in a typical case.

4.5.3 The PID H∞ RPM Regulator with Shaft Torque Feedforward

In Section 4.4.3 the gains of a general PID control law were calculated, so that the H∞-norm of the closed-loop transfer function from propeller load disturbance to shaft rpm variation does not exceed a specified value. In the current section, the modified gains of the PID control law will be determined, taking into account that in practical applications it is preferable to implement the D-term using the shaft torque feedforward instead of direct differentiation of the rpm signal.

The starting point will be to assume certain known values for the P, I and D gains, K_{P0}, K_{I0} and K_{D0} respectively. These values have been calculated according to the equations presented in Section 4.4.3, so that the closed-loop transfer function obtains damping ratio $\zeta = 1.0$, as well as appropriate values for steady-state gain K and natural underdamped frequency ω_n for meeting the H∞-norm requirement mentioned earlier. Additionally, ω_n should fulfil the requirements for relative stability (see Section 4.4.3) and robustness against neglected dynamics (see Section 4.4.4), as well as to guarantee that closed-loop transfer function $\tilde{G}_c(s)$ from the propulsion plant disturbance to the control action signal is minimum-phase (see Section 4.4.3).

Using the above-mentioned starting values of the P, I and D gains, the gains of the modified control law, K_P, K_I and K_{FF}, where the D-term has been replaced by the shaft, torque feedforward will be now calculated.

As already mentioned, the PID speed governor generates a control action (fuel index deviation $f_R(t)$ from the steady-state fuel index value F_{R0} at the specified engine loading) based only on the shaft rpm deviation signal $n(t)$ (or equivalently to speed error $e(t)$, by inverting the sign of all constants) according to the following integro-differential relationship:

$$f_R(t) = K_P \cdot n(t) + K_I \cdot \int_{t_0}^{t} n(\xi)d\xi + K_D \cdot \frac{d}{dt}n(t)$$

$$\Updownarrow$$

$$f_R(s) = \left(K_P + \frac{K_I}{s} + K_D s \right) \cdot n(s) \tag{4.97}$$

As has been explained in Section 4.5.2, the cycle-mean rpm rate of change can be estimated by:

$$\dot{N}(t) = \frac{Q_E(t) - Q_{shaft}(t)}{I_E} = \frac{C \cdot F_R(t) - Q_{fr} - Q_{shaft}(t)}{I_E} \tag{4.98}$$

For the rpm regulation problem of the marine plant, no change of operating point is assumed. Therefore:

$$\dot{n}(t) \equiv \frac{d}{dt}n(t) \triangleq \frac{d}{dt}\left(N(t) - N_0\right) = \dot{N}(t) \tag{4.99}$$

Following the quasi-steady modelling assumption, shaft torque Q_{shaft} is analysed to a steady-state value plus perturbation signal, Q_0 and $r(t)$, according to the following:

$$Q_{shaft}(t) = Q_0 + r(t) \tag{4.100}$$

For the speed regulation problem the steady-state value Q_0 is assumed not to change. Furthermore, in steady-state it holds that:

$$\left. \begin{array}{l} \dot{N}_E = \dot{N}_P = 0 \\ N_{E0} = N_{P0} = N_0 \end{array} \right\} \Rightarrow$$

$$Q_0 = Q_{E0} = Q_{P0} \Rightarrow Q_0 = C \cdot F_{R0} - Q_{fr} = K_{Q0} \cdot N_0^2 \tag{4.101}$$

Taking into account the above, the equation for the cycle-mean estimate of shaft rpm rate of change is transformed to the following:

$$\dot{n}(t) = \frac{C \cdot f_R(t) - r(t)}{I_E} \tag{4.102}$$

By substituting the above equation for $\dot{n}(t)$ in the PID control law, we finally obtain that:

$$f_R(t) = \frac{I_E K_{P0}}{I_E - CK_{D0}} \cdot n(t) + \frac{I_E K_{I0}}{I_E - CK_{D0}} \cdot \int_{t_0}^{t} n(\xi) d\xi - \frac{K_{D0}}{I_E - CK_{D0}} \cdot r(t) \quad (4.103)$$

Therefore, the gains of the modified control law, including shaft rpm feedback and shaft torque feedforward for the implementation of D-term (Hinf PI+FF regulator), are calculated from the Hinf PID regulator gains according to the following:

$$K_P = \frac{I_E}{I_E - CK_{D0}} \cdot K_{P0},\ K_I = \frac{I_E}{I_E - CK_{D0}} \cdot K_{I0},\ K_{FF} = -\frac{K_{D0}}{I_E - CK_{D0}} \quad (4.104)$$

where evidently:

$$f_R(t) = K_P \cdot n(t) + K_I \cdot \int_{t_0}^{t} n(\xi) d\xi + K_{FF} \cdot r(t) \quad (4.105)$$

As $K_{D0} < 0 \Rightarrow I_E - CK_{D0} > I_E$, it holds that:

$$K_P < K_{P0} \text{ and } K_I < K_{I0} \quad (4.106)$$

The reduction in the P and I gains of the control law can be regarded as a positive side effect of shaft torque signal exploitation. Indeed, the (rated) fuel index value is limited in the range 0–.100%. Therefore, smaller P and I gains are translated to increased margin until saturation of the control action occurs; note that saturation is a non-linear effect and therefore should be avoided. Finally, note that the feedforward gain K_{FF} has the opposite sign in relation to K_{D0}, as increase in shaft torque results in engine deceleration and *vice versa*.

4.5.4 Typical Case Numerical Investigation

In Section 4.4.5 the speed regulation of the propulsion powerplant of containership "Shanghai Express" was examined. There, it was shown that only the Hinf PID regulator can meet the H∞-norm requirement and at the same time provide robustness against the neglected dynamical terms. Here, the PID control law is substituted with the Hinf PI+FF scheme; it is also noted that onboard "Shanghai Express" the shaft torque signal is available for feedback, as there exists a torque meter installation on the engine–propeller shafting system.

The gains of the Hinf PID control law are given in Table 4.2 and have been translated to the gains of the Hinf PI+FF scheme shown in Table 4.3.

Table 4.2 Hinf PID regulator gains for "Shanghai Express" powerplant

	$(-K_P)$ (%index/rpm)	$(-K_I)$ (%index/rpm/s)	$(-K_D)$ (%index · s/rpm)
Hinf PID	13.19	11.67	2.53

Table 4.3 Hinf PI+FF regulator gains for "Shanghai Express" powerplant

	$(-K_P)$ (%index/rpm)	$(-K_I)$ (%index/rpm/s)	$(-K_D)$ (%index · s/rpm)
Hinf PI+FF	4.54	4.01	0.0226

For the above calculation the assumption $I_E \approx I_P \approx I/2$ has been used, instead of the exact value of engine inertia I_E.

The Hinf PI+FF control scheme has been validated in simulation using the block diagram shown in Figure 4.13 for the powerplant of "Shanghai Express". In this block diagram, shaft torque has been calculated using the quasi-steady relationship:

$$2 \cdot Q_{shaft} = Q_E + Q_P \Rightarrow 2 \cdot r(t) = Q_E(t) + Q_P(t) - 2 \cdot Q_0 \qquad (4.107)$$

Therefore, in simulation, signal $(2 \cdot r(t))$ is generated based on the current values of engine torque delivery and propeller torque demand; then, it is propagated to contribute to the fuel index value through a feedforward gain with value $(K_{FF}/2)$.

Furthermore, as can be seen in Figure 4.13, the linearised version of the propeller law has been used in the simulation. Finally, the block of neglected dynamics is identical to that used for validation of the Hinf PI and PID regulators previously, including a turbocharging/combustion term with $\tau_{TC} = 0.25$ s and an actuator term with time constant $\tau_{act} = 0.10$ s.

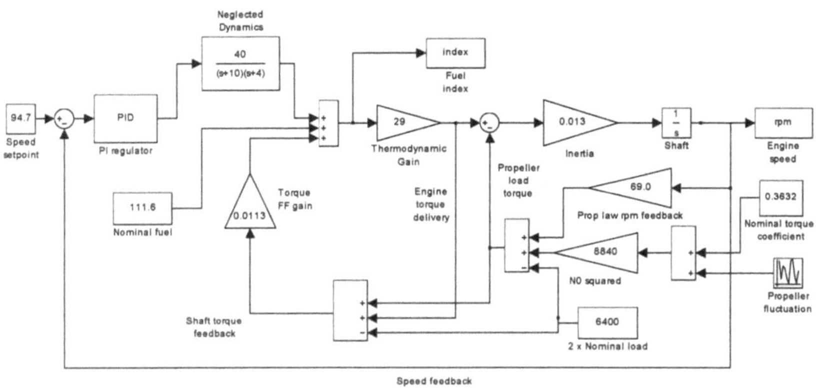

Figure 4.13 Hinf PI+FF scheme for "Shanghai Express" plant used for simulations

The response of the Hinf PI+FF scheme in comparison with the Hinf PID regulator is given in Figure 4.14, where the response of the closed-loop plant in terms of shaft rpm and fuel index to a torque coefficient (k_Q) step of 0.05 kN m/rpm^2 at $t = 5$ s is presented.

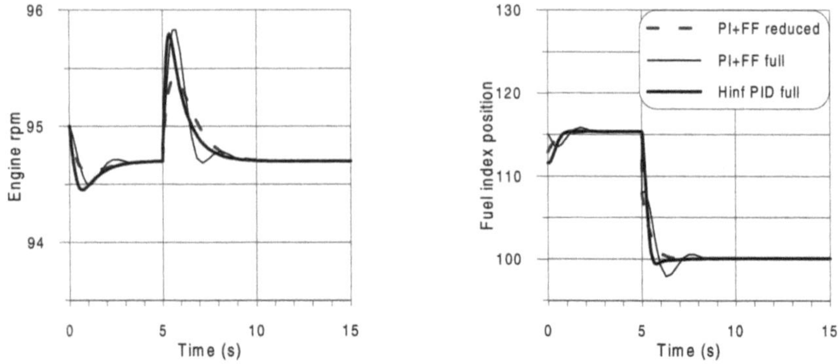

Figure 4.14 Response of the Hinf PI+FF scheme compared to Hinf PID regulator

At first, it can be readily observed that the Hinf PI+FF performance, when coupled to the reduced-order transfer function of the marine plant, is identical to that of the Hinf PID regulator. Then the response of the full-order transfer function of the marine plant with both the Hinf PID and PI+FF schemes is plotted. As can be seen, although major features of the response (*e.g.* overspeed, settling time, *etc.*) are the same, there exist slight differences in the shape of the response. For example, the Hinf PI+FF response exhibits some small ripple and, in effect, some undershoot in fuel index, whereas the Hinf PID does not. This can be attributed to the fact that the approximation of rpm derivative, using the shaft torque signal, is valid in the low-frequency range; however, a step excitation signal also gives a significant high-frequency spectral content. Therefore, the response of the plant carries some undesired high-frequency components, which, however, do not make the Hinf PI+FF scheme unacceptable.

Therefore, it can be concluded that the Hinf PI+FF scheme can be used shipboard for enhanced propeller disturbance rejection. The gains of the control law can be determined using those of the Hinf PID regulator; however, as is often the case, in PI(D) control practice fine tuning may be required, when the controller is coupled to the actual plant, for further improvement in performance and adaptation to the actual operating conditions.

4.6 Summary

A method for robust PID speed governor design method is proposed. The method is based on the notion of H_∞-norm and aims to cover the gap existing in the marine engine control field, as far as rejection of severe and rather fast load disturbance is concerned. This gap leads, quite often in marine seagoing practice, to deliberately decrease engine rating in order to avoid potential engine overloading due to overspeed. Therefore, loop-shaping with the PID control law is investigated so that adequate disturbance attenuation is achieved. The rejection achievement is quantified by the H_∞-norm of the closed-loop transfer function from propeller disturbance to shaft rpm. On the other hand, severe load disturbance rejection requirements lead to making the closed-loop system comparatively fast, meaning that neglected dynamics may start to become important. In the case of marine propulsion systems, by employing the full-order open-loop transfer function, obtained in Chapter 3, it has been proven that only a PID controller can meet the robustness requirements. Therefore, in order to avoid real-time differentiation of the rpm feedback signal, the technique of disturbance feedforward has been examined. In the case of marine plants, a control scheme has been proposed incorporating two feedback signals, namely shaft rpm and torque, instead of shaft rpm only. In effect, a regulator, based on the tuning of the H_∞ PID speed governor, but with the shaft torque signal substituting the D-term, is finally proposed.

CHAPTER 5

STATE-SPACE DESCRIPTION OF THE MARINE PLANT

5.1 Introduction

5.1.1 Overview of the Approach

As already mentioned, the first step towards the application of advanced control schemes, ensuring both disturbance rejection and robustness against the various forms of uncertainty, is the construction of a state-space model for the operation of the plant. This chapter is concerned with the development of a non-linear state-space model that can depict the marine engine–turbocharger dynamical interaction and operation, on the one hand, and integrate the inherent physical uncertainty and disturbance on the other hand. The state-space model is derived using the non-linear mapping abilities of artificial neural nets. Although neural nets have been employed in the past for the description of engine physicochemical processes, especially in the automotive industry, the approach presented here is different for two major reasons. First, the approach presented does not aim to fill some "gap of understanding" in an otherwise full physical modelling picture. Quite the opposite, it employs the full picture of the cycle-mean, quasi-steady, thermodynamic model of Chapter 2 in order to bypass it, because it requires the numerical solution of a non-linear, perplexed algebraic system of equations. Moreover, the neural nets are treated rather as "mathematical objects" than as part of a global approach to intelligent powerplant modelling and control. Therefore, the mathematical expressions corresponding to the typical feedforward neural net structure with one hidden layer are manipulated analytically in order to derive a linearised, yet uncertain, perturbation state-space model.

In contrast to the case of automotive engines, which in many cases are not turbocharged, the experimental determination of the state equations or transfer function is not possible for marine propulsion engines. This is due to a number of reasons. First of all the large power output and physical size of marine engines make the construction of testbed facilities a costly task, whilst the deployment of a new engine in a testing facility is also time consuming. No prototype marine engines are usually available, and the performance of the actual shipboard plant is, in most cases, much different than the one recorded in possible engine tests at shore facilities.

The alternative way of constructing and identifying the marine propulsion powerplant state-space equations is to derive them from physical principles. The complexity of the processes and phenomena, however, requires the use of detailed physical numerical simulation models incorporating the principles and quantitative relationships dictated by thermodynamics [39,40]. As already demonstrated in Chapter 2, the mathematical interdependence between the various plant variables is neither explicit nor direct. Actually, these variables do appear in a set of non-linear and perplexed algebraic relations. Assuming the standard form of state-space equations:

$$\frac{d\vec{x}}{dt} = \vec{f}(\vec{x},\vec{u}) \tag{5.1}$$

the mapping $\vec{f}$ involved cannot be readily formulated for marine propulsion powerplants. The numerical iterative solution of the algebraic part of the physical (thermodynamic) model, such as the one in Chapter 2, though, can provide a grid of points for which the value of $\vec{f}$ is known. Then suitably "dimensioned" neural nets can be trained, using the generated grid, to approximate $\vec{f}$ with desired accuracy [41,42]. The neural nets produced by the above procedure depict mapping $\vec{f}$ with required accuracy and in explicit, closed mathematical form.

Then, an appropriate decomposition scheme can be applied to the plant state equations implying the partitioning of control action and controllers for the controller synthesis stage. Actually, the decomposition produces two discrete open-loop plant models: the Non-linear Nominal Model (N^2M) and the linear Uncertain Perturbation Model (UPM). The objective is to separate the control action, which in the case of marine propulsion plants is the engine fuel index, into two major parts: the steady-state fuelling demand and the perturbation control action that it is aimed to minimise the effect of propeller disturbance. The fuelling demand represents the propulsion power requirements of the ship "as is", *i.e.* under the specific loading (trim and draft), weather and sea conditions, as well as due to ageing effects, *e.g.* hull and propeller fouling. Short-term feedback control is not necessary for fuelling demand, as will be shown in the next chapter. However, for efficient ship management purposes, slow, long-term adjustment of this portion of the control action can be employed. Adjustment can be based on monitoring the conditions mentioned above in conjunction with engine performance indices. Furthermore, it can be done either manually or automatically. Automatic adjustment of fuel index offset value can be effectuated by intelligent engine management systems in order to minimise fuel consumption or pollutant emissions or an appropriately weighted combination of both.

On the other hand, feedback controls are required in the case of UPM, if closed-loop transient performance requirements need to be met, *e.g.* when the propeller torque fluctuations are significant. The extensive results of control theory of linear systems can be applied in this case. However, the presence of uncertainty has to be taken into account. A major concern is the parametric uncertainty introduced due to the linearisation procedure, because the values of the partial derivatives depend upon the equilibrium point at which they are calculated, whilst another is due to the inherent uncertainty of physical parameters. In that respect, development of robust feedback control of the UPM state-space equations is a core subject of this work.

The first step towards this objective, however, is the formulation of an open-loop state-space model and a transfer function matrix (open-loop, as well) for the marine plant. A neural, state-space model will be deployed on the basis of the thermodynamic engine description given in Chapter 2. Finally, the transfer function matrix obtained from the state-space model will be deduced and compared to the one obtained empirically.

5.1.2 Mathematical Formulation and Notation

It is widely accepted that advanced control methods rely on state-feedback for improving the closed-loop dynamical behaviour. The state of the controlled system includes all the variables of interest that appear in the descriptive dynamical equations. The general form of the state equations of a system is:

$$\begin{aligned} \dot{\mathrm{x}} &= \mathrm{f}(\mathrm{x},\mathrm{u}) \\ \mathrm{y} &= \mathrm{h}(\mathrm{x}) \end{aligned} \tag{5.2}$$

where $\mathrm{x}^{\mathrm{T}}=[x_1\ x_2\ \ldots\ x_n]$ is the n-dimensional state column vector, including all the dynamical variables of interest, $\mathrm{u}^{\mathrm{T}}=[u_1\ u_2\ \ldots\ u_m]$ is the m-dimensional column vector of control actions applied to the system and $\mathrm{y}^{\mathrm{T}}=[y_1\ y_2\ \ldots\ y_p]$ is the p-dimensional vector of the system measured outputs. Function $\mathrm{f}(\mathrm{x},\mathrm{u}) = \mathrm{f}(x_1,x_2,\ldots,x_n,u_1,u_2,\ldots,u_m)$ is an $\mathbb{R}^n$-valued function (mapping) defined on a subset of $\mathbb{R}^{(n+m)}$. Function $\mathrm{h}(\mathrm{x}) = \mathrm{h}(x_1,x_2,\ldots,x_n)$ is an $\mathbb{R}^p$-valued function (mapping) defined on a subset of $\mathbb{R}^n$. As is obvious, italic letters denote scalar variables and lower-case, non-italic ones vector quantities. Matrices will be denoted using capital, non-italic letters or lower-case, non-italic ones using the symbol $\ddot{}$ above them.

The points of space $\mathbb{R}^{(n+m)}$ (*i.e.* $(n+m)$-ads) at which:

$$\dot{\mathrm{x}} = \mathrm{f}(\mathrm{x},\mathrm{u}) = \vec{0} \tag{5.3}$$

will be called equilibrium points in the rest of this text. As will be seen later on, for marine plants these points are the steady-state operating points, around which regulation for disturbance rejection is required.

Control of systems with the above generic form of non-linear state equations has demonstrated significant advances in recent years. For example, in [43] the analysis of a wide class of non-linear systems is covered in great detail; then, a variety of control problems is addressed, including feedback linearisation, disturbance decoupling, output regulation, *etc.* The state equations of the systems addressed in work [43] give rise to the following "control-action-affine" form:

$$\dot{\mathrm{x}} = \mathrm{f}(\mathrm{x}) + \sum_{i=1}^{m} \mathrm{g}_i(\mathrm{x}) \cdot u_i \equiv \mathrm{f}(\mathrm{x}) + \vec{\mathrm{g}}(\mathrm{x}) \cdot \mathrm{u} \qquad (5.4)$$
$$\mathrm{y} = \mathrm{h}(\mathrm{x})$$

where $\mathrm{g}_i(\mathrm{x}) = \mathrm{g}_i(x_1, x_2, \ldots, x_n)$ are $\mathbb{R}^n$-valued functions defined on a subset of $\mathbb{R}^n$. Finally, $\vec{\mathrm{g}}(\mathrm{x})$ is an $(n \times m)$-matrix function defined as:

$$\vec{\mathrm{g}}(\mathrm{x}) = [\mathrm{g}_1(\mathrm{x})\ \mathrm{g}_2(\mathrm{x})\ \ldots\ \mathrm{g}_m(\mathrm{x})] \qquad (5.5)$$

Another approach, however, to the control problem of non-linear plants is linearisation. Linearisation of the state equations is performed around equilibrium points. For a given equilibrium point, *e.g.* $\left[\mathrm{x}_0^{\mathrm{T}}\ \mathrm{u}_0^{\mathrm{T}}\right]^{\mathrm{T}} \in \mathbb{R}^{(n+m)}$, linearisation of $\mathrm{f}(\mathrm{x},\mathrm{u})$ can be performed as follows:

$$\mathrm{f}(\mathrm{x},\mathrm{u}) \approx \mathrm{f}(\mathrm{x}_0,\mathrm{u}_0) + \sum_{i=1}^{n} \left.\frac{\partial \mathrm{f}}{\partial x_i}\right|_{(\mathrm{x}_0,\mathrm{u}_0)} \cdot (x_i - x_{i0}) + \sum_{i=1}^{m} \left.\frac{\partial \mathrm{f}}{\partial u_i}\right|_{(\mathrm{x}_0,\mathrm{u}_0)} \cdot (u_i - u_{i0}) \qquad (5.6)$$

In the above, the n-dimensional partial derivative vectors are calculated by the following definitions, provided that $\mathrm{f}(\mathrm{x},\mathrm{u})$ is known:

$$\left.\frac{\partial \mathrm{f}}{\partial x_i}\right|_{(\mathrm{x}_0,\mathrm{u}_0)} \triangleq \left[\left.\frac{\partial}{\partial x_i} f_1(\mathrm{x},\mathrm{u})\right|_{(\mathrm{x}_0,\mathrm{u}_0)} \ \left.\frac{\partial}{\partial x_i} f_2(\mathrm{x},\mathrm{u})\right|_{(\mathrm{x}_0,\mathrm{u}_0)} \ \cdots \ \left.\frac{\partial}{\partial x_i} f_n(\mathrm{x},\mathrm{u})\right|_{(\mathrm{x}_0,\mathrm{u}_0)} \right]^T, \ i = 1, \ldots, n \qquad (5.7)$$

$$\left.\frac{\partial \mathrm{f}}{\partial u_i}\right|_{(\mathrm{x}_0,\mathrm{u}_0)} \triangleq \left[\left.\frac{\partial}{\partial u_i} f_1(\mathrm{x},\mathrm{u})\right|_{(\mathrm{x}_0,\mathrm{u}_0)} \ \left.\frac{\partial}{\partial u_i} f_2(\mathrm{x},\mathrm{u})\right|_{(\mathrm{x}_0,\mathrm{u}_0)} \ \cdots \ \left.\frac{\partial}{\partial u_i} f_n(\mathrm{x},\mathrm{u})\right|_{(\mathrm{x}_0,\mathrm{u}_0)} \right]^T, i = 1, \ldots, m \qquad (5.8)$$

In the sequel, the following definitions for the $(n \times n)$ and $(n \times m)$ "derivative" matrices of the non-linear system are adopted:

$$\vec{\mathrm{f}}'_{\mathrm{x}}(\mathrm{x}_0,\mathrm{u}_0) \triangleq \left[\left.\frac{\partial \mathrm{f}}{\partial x_1}\right|_{(\mathrm{x}_0,\mathrm{u}_0)} \ \left.\frac{\partial \mathrm{f}}{\partial x_2}\right|_{(\mathrm{x}_0,\mathrm{u}_0)} \ \cdots \ \left.\frac{\partial \mathrm{f}}{\partial x_n}\right|_{(\mathrm{x}_0,\mathrm{u}_0)} \right] \qquad (5.9)$$

$$\vec{\mathrm{f}}'_{\mathrm{u}}(\mathrm{x}_0,\mathrm{u}_0) \triangleq \left[\left.\frac{\partial \mathrm{f}}{\partial u_1}\right|_{(\mathrm{x}_0,\mathrm{u}_0)} \ \left.\frac{\partial \mathrm{f}}{\partial u_2}\right|_{(\mathrm{x}_0,\mathrm{u}_0)} \ \cdots \ \left.\frac{\partial \mathrm{f}}{\partial u_m}\right|_{(\mathrm{x}_0,\mathrm{u}_0)} \right] \qquad (5.10)$$

Also, the perturbation (around the equilibrium point $\left[\mathrm{x}_0^{\mathrm{T}}\ \mathrm{u}_0^{\mathrm{T}}\right]^{\mathrm{T}}$) state and control action vectors, $\delta\mathrm{x}$ and $\delta\mathrm{u}$ respectively, can be defined as following:

$$\delta\mathrm{x} = \mathrm{x} - \mathrm{x}_0 \ \text{and} \ \delta\mathrm{u} = \mathrm{u} - \mathrm{u}_0 \qquad (5.11)$$

Taking into account the fact that $\mathrm{f}(\mathrm{x}_0,\mathrm{u}_0) = \vec{0}$, the linearisation equation takes the following form, in an appropriately small vicinity of the equilibrium point $\left[\mathrm{x}_0^{\mathrm{T}}\ \mathrm{u}_0^{\mathrm{T}}\right]^{\mathrm{T}}$:

$$f(x,u) \approx \overleftrightarrow{f}'_x(x_0,u_0)\cdot\delta x + \overleftrightarrow{f}'_u(x_0,u_0)\cdot\delta u \tag{5.12}$$

Note that if the equilibrium point is not assumed to change, either deliberately or unwillingly, then the following holds:

$$\delta\dot{x} = \frac{d}{dt}(x - x_0) = \dot{x} \tag{5.13}$$

Therefore, the non-linear state equation:

$$\dot{x} = f(x,u)$$

is reduced to the following linear one:

$$\delta\dot{x} = \overleftrightarrow{f}'_x(x_0,u_0)\cdot\delta x + \overleftrightarrow{f}'_u(x_0,u_0)\cdot\delta u \tag{5.14}$$

where, obviously the state and control action vectors, x and u respectively, have been replaced by their perturbation counterparts, δx and δu. Therefore, and due to the availability of a large number of theoretical tools and techniques on the analysis and control of linear systems, the control problem of the non-linear system is greatly simplified and facilitated. Note that in the above analysis the second equation for the calculation of measured outputs y from the system states x has not been included; the extension of the lineariståion method is straightforward and will not be addressed any further in this text.

However, there is a penalty for the simplification provided by system linearisation. Both the perturbation plant model and the controls developed give rise to a "localised" character due to the fact that the linearisation of f(x,u) is valid only in a small vicinity of the specific equilibrium point $\left[x_0^T\ u_0^T\right]^T$ under consideration.

This difficulty, though, can be dealt with by considering a family of linearised plant models, instead of a single one that approximates the actual non-linear model only in the vicinity of a specific $\left[x_0^T\ u_0^T\right]^T$. To be specific, the modelling difficulty, when using linear models, originates from the fact that the values of each of the two derivative matrices, namely $\overleftrightarrow{f}'_x$ and $\overleftrightarrow{f}'_u$, are not constant (otherwise the system would be linear in the first place), but depend upon the operating point around which the linearisation takes place. In many cases, though, as it is for marine plants, the derivative matrices can be bounded in a region of $\mathbb{R}^{n\times n}$ and $\mathbb{R}^{n\times m}$, for all steady-state (equilibrium) operating points $\left[x_0^T\ u_0^T\right]^T$ in the range of practical interest, as follows:

$$\overleftrightarrow{f}_{x,LO} \le \overleftrightarrow{f}'_x(x_0,u_0) \le \overleftrightarrow{f}_{x,HI} \text{ and } \overleftrightarrow{f}_{u,LO} \le \overleftrightarrow{f}'_u(x_0,u_0) \le \overleftrightarrow{f}_{u,HI} \tag{5.15}$$

In other words, the nonlinearity of the original plant model is "translated" to parametric uncertainty of the derivative matrices appearing in the linearised perturbation models. For linear systems with parametric uncertainty as above, the theory of linear robust control can, if certain conditions are met, provide a unique state or output feedback controller that guarantees plant transient performance, with respect to certain specifications for disturbance rejection in the time, or more often, in the complex frequency domain. Therefore, the problem of non-linear control is converted to a problem of controller synthesis for an entire family of linear systems.

The problem of robust control of uncertain linear systems is addressed in great extent in [44] and in [45]. Of course, in order to apply the results available in these two or other similar studies, *e.g.* [46], a description of the system with non-linear state-equations is required. Then, after the procedure of linearisation of the non-linear state equations has been carried out, the bounds for the derivative matrices can be determined. As shown in Chapter 2, the marine plant dynamical equations include one for the engine–propeller shaft and one for the turbocharger shaft. They are repeated here for the sake of completeness:

$$\dot{N}(t) \equiv \dot{N}_E(t) = \frac{Q_E - Q_L}{I} \text{ and } \dot{N}_{TC}(t) = \frac{Q_T + Q_C}{I_{TC}} \tag{5.16}$$

Therefore, the state vector of the marine plant is:

$$\mathrm{x} = \begin{bmatrix} N_E \\ N_{TC} \end{bmatrix} \tag{5.17}$$

Moreover, for conventional marine engines, such as those addressed in the present text, the only available variable for feedback control is the fuel amount injected in each cylinder per cycle, as expressed by the, usually rated (dimensionless), fuel index (rack) value. Therefore:

$$\mathrm{u} \equiv u = F_R \tag{5.18}$$

As has already been argued, propeller load torque depends mainly on propeller rotational speed (rpm). Taking into account that the engine–propeller shaft dynamics of order higher than one can be neglected:

- engine load torque coincides with propeller load torque, and
- propeller speed coincides with shaft and engine speed.

Therefore, propeller law dictates the functional dependence of load torque on one of the plant's state variables. However, in order to formulate the non-linear state-space equations of the marine plant it is required to derive direct mathematical and differentiable expressions for engine, turbine and compressor torque, as well. In brief, functions of the following form are sought after:

$$Q_E(t) = Q_E(N(t), N_{TC}(t), F_R(t)) \text{ and } Q_{TC}(t) = Q_{TC}(N(t), N_{TC}(t), F_R(t)) \tag{5.19}$$

where $Q_{TC}(t) = Q_T(t) + Q_C(t)$ is the total accelerating or decelerating torque applied to the turbocharger shaft, combining turbine and compressor torque. Note that for all variables (torque, rpm and fuel index), the thermodynamic cycle mean value is the one appearing in the above expressions. This is in agreement with the assumption of the engine thermodynamic model developed in Chapter 2.

In Section 5.2 a method employing artificial neural nets for approximating the above functions for the engine and turbocharger torque variables, namely $Q_E(N, N_{TC}, F_R)$ and $Q_{TC}(N, N_{TC}, F_R)$ respectively, is presented. The method utilises steady-state and performance data derived from the solution of the quasi-steady, cycle-mean, thermodynamic engine model of Chapter 2. For convenience the following vector (triad) of inputs to the neural nets is defined:

$$\mathrm{X} = [N \quad N_{TC} \quad F_R]^{\mathrm{T}} \tag{5.20}$$

which "packs" the state vector x and scalar control action u of the marine plant.

Before concluding this introductory section, it should be noted that the neural torque approximators are independent from any engine thermodynamic model. The sole necessity for such a model originates of the fact that a significant volume of data is required for the calibration (training in the neural nets terminology) of the approximators. The data must cover both equilibrium (steady-state), as well as dynamic transient operating conditions. If such data are available in some other manner, *e.g.* from experimental data series or from detailed emptying-and-filling thermodynamic simulation codes like MoTher (presented previously), the use of the quasi-steady, cycle-mean model is redundant.

Therefore, the value of the engine model of Chapter 2 for control purposes is to provide significant insight to the engine physicochemical processes. Specifically, the thermodynamic model was the one that indicated that the order of the state-space equations is two, if cycle-mean approximation is required. For example, on the other hand, MoTher provides in-cycle variations of the plant thermodynamic and mechanical variables, but with a significant number of additional differential equations for depicting the dynamical processes of volume emptying-and-filling during each cycle. Therefore, if this were the physical model of reference, it would have been required to use a state-space description of a much higher order. However, here it worth remembering that the sole control action available for conventional marine plants is the fuel index value, which has a once-per-cycle effect. Therefore, the in-cycle temporal resolution of MoTher and similar models is not required and makes controller synthesis more complicated.

The other option for training the neural approximators is to employ measured torque data series. In the first place, though, these series are extremely hard to find, especially in the case of marine plants. Moreover, turbocharger torque is not easily measured. Furthermore, the algebraic part of the thermodynamic model can be solved for arbitrary values of the input vector X. This allows estimating the values of torque at situations away from thermodynamic equilibrium, which is the case in extreme transients. Therefore, the bounds for the derivative matrices can be determined more accurately when using the predictions of the quasi-steady model. On the other hand, experimental data series, especially for turbocharger torque, covering such large operating regions cannot include such extreme situations, due to safety precautions. Therefore, the use of comprehensive physical models becomes indispensable if the bounds of the uncertain derivative matrices involved in the linearisation procedure are to be conservatively determined.

5.2 The Neural Torque Approximators

5.2.1 Configuration of the Approximators

As early as in Chapter 2, the thermodynamic, quasi-steady, cycle-mean marine engine model derived from physical principles has been separated into a dynamical and an algebraic part. The first one dictated the state differential equations, whilst the second brought forth the non-linear character of the processes involved in engine power/torque generation. However, as already seen, the algebraic part of the model consists of a non-linear, perplexed set of equations. The solution of this system cannot proceed otherwise but numerically. The outcome, therefore, is a fine grid of points, defined by values for the triad (N, N_{TC}, F_R), for which the values of engine, turbine and compressor torque, Q_E, Q_T and Q_C respectively, are calculated. The three torque maps generated, however, do not provide the explicit functional dependence of the torque variables upon state variables (engine and turbo rpm) and the control action (fuel index). In effect, although they can be used, in conjunction with an appropriate interpolation scheme for reducing the computational effort required for the transient simulation of a marine engine (as the algebraic part may be bypassed), they cannot be employed for controller synthesis with advanced, analytical theoretical tools.

Alternatively, curve-fitting techniques can be used for the approximation of the torque maps. For example in [47] the following expression for engine torque delivery Q_E is given:

$$Q_E(N, F_R) = \alpha_{\dagger} \cdot F_R + \beta_{\dagger} + \gamma_{\dagger} \cdot N + \delta_{\dagger} \cdot N^2 \tag{5.21}$$

Coefficients of the above expression can be determined by least-squares curve-fitting of either experimental data series, if available, or from the engine torque map generated by the quasi-steady, cycle-mean thermodynamic model. The problem with this approach is that the functions to which experimental or model-predicted data can be fit are not known, especially for turbine and compressor torque, Q_T and Q_C respectively. Furthermore, if an estimated guess is possible for one or more torque functions of X, *e.g.* deduced from inspection of the curves, its applicability may not be of general validity for any marine engine. Therefore, what is sought is an automated procedure that allows approximation of the three torque hypersurfaces (subsets of $\mathbb{R}^4$), with required accuracy. The torque hypersurfaces are parameterised by the three-dimensional input vector X and generated by the thermodynamic model.

From the perspective of overcoming the difficulty of torque approximation, artificial neural nets have been considered. The major attractive feature of three (or more) layered, fully connected, feed-forward neural networks is their ability to depict non-linear mappings in a closed, standard and analytical mathematical form, which is extremely valuable for the application of modern control theory. Moreover, neural networks do not require manual tuning; indeed, by using methods inspired from optimisation theory, automated calibration of the standard, mathematical form implied by the net to the targeted mapping is straightforward. Major theoretical results from the discipline of neural nets cover approximation accuracy, as well as training methodology and convergence aspects. To be specific, under the assumption that functions $Q_E(N, N_{TC}, F_R)$ and $Q_{TC}(N, N_{TC}, F_R)$ are continuous, then appropriately sized neural nets can be used for the approximation of the torque maps generated, according to the theorems of Kolmogorov (1957) and Hecht-Nielsen (1987) [48]. Kolmogorov's result states that any continuous (scalar) function (map) $\mathrm{F}:\Im^{\nu} \rightarrow \Im$ where $\Im = [0,1]$ of ν variables can be represented by a three-layer, feed-forward neural net with a $(2\nu+1)$-neurons hidden layer. Specifically, the theorem guarantees that F can be put in the form:

$$\mathrm{F}(\vec{X}) = \sum_{j=1}^{2\nu+1} \chi_j \Big(\sum_{i=1}^{\nu} \psi_{ij}(X_i)\Big), \text{ where } \vec{X} = \begin{bmatrix} X_1 & X_2 & \cdots & X_\nu \end{bmatrix}^{\mathrm{T}} \tag{5.22}$$

where $\chi_j(\xi),\ j = 1,\ldots,(2\nu+1)$ and $\psi_{ij}(\xi),\ i = 1,\ldots\nu,\ j = 1,\ldots,(2\nu+1)$ are continuous functions of one variable (ξ) and $\psi_{ij}(\xi)$ are monotonic functions that do not depend on the specific mapping F. In some sense, Kolmogorov's result resembles the expansion of periodic functions into Fourier series. The main difference is that Fourier series are linear combinations (weighted sums) of sines and/or cosines, with frequency multiples of the principal harmonic. On the other hand, Kolmogrov's "expansion" is a somehow "non-linear" combination of the selected basis functions $\psi_{ij}(\xi),\ i = 1,\ldots\nu,\ j = 1,\ldots,(2\nu+1)$, as they have to "propagate" through the next layer's functions $\chi_j(\xi),\ j = 1,\ldots,(2\nu+1)$.

In effect, Kolmogorov's expansion for any non-linear mapping F implies the fully connected feed-forward layered architecture with weights shown in Figure 5.1.

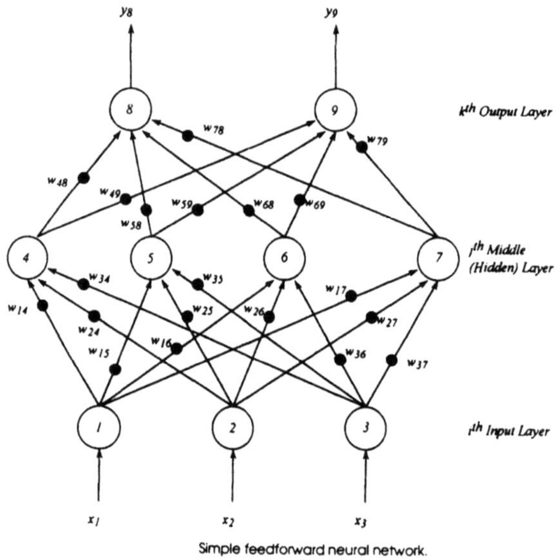

Figure 5.1 Typical architecture of fully connected, feed-forward three-layered neural net

In practice, the following additional architectural features are used for feedforward nets [49]:

- The outputs of the net are also normalised in the "unit" interval $\Im = [0,1]$, although Kolmogorov's theoretical result requires normalisation of the inputs only.
- The input layer is not a neural computing element, *i.e.* the nodes do not have input weights and activation functions assigned to them, *i.e.* $\psi_{ij}(X_i) = X_i$.
- As indicated by Kolmogorov's result, there is only one neuron at the output layer, summing up the outcome of the hidden layer non-linear neurons.
- The activation function for the hidden layer nodes $\chi_j(\xi)$ can be any monotonically increasing function that is everywhere differentiable. Usually it is the logistic sigmoid, *i.e.*:

$$\Phi(\xi) = \frac{1}{1+e^{-\xi}} = \frac{e^{\xi}}{1+e^{\xi}} \tag{5.23}$$

- Biases, are added to the output of each hidden and input layer node. Biases are used in order to regulate the net input to each unit (node) [48].

In the case of marine plants neural nets have been employed for the approximation of the torque $\mathbb{R}^4$-hypersufaces. The torque maps derived from the thermodynamic model are parameterised by the triad (N, N_{TC}, F_R). Therefore, the number of nodes for the hidden layer is chosen to be seven, in accordance with Kolmogorov's result. In conclusion, the neural approximators for engine torque $Q_E(\mathrm{X})$ and turbocharger total torque $Q_{TC}(\mathrm{X}) = Q_T(\mathrm{X}) + Q_C(\mathrm{X})$ take the following standard mathematical functional form:

$$Q_E(\mathrm{X}) = \mathrm{Q_{Emax}} \cdot \{w_{00} + \mathrm{W}_0 \cdot \Phi(\mathrm{W}_b + \mathrm{W} \cdot \mathrm{X})\} \tag{5.24}$$

$$Q_{TC}(\mathrm{X}) = \mathrm{Q_{Tmax}} \cdot \{v_{00} + \mathrm{V}_0 \cdot \Phi(\mathrm{V}_b + \mathrm{V} \cdot \mathrm{X})\} + \mathrm{Q_{Cmax}} \tag{5.25}$$

In the above, the vector form of the logistic sigmoid function is defined by the following:

$$\Phi(\vec{\xi}) = [\Phi(\xi_1) \quad \Phi(\xi_2) \quad \cdots \quad \Phi(\xi_7)]^{\mathrm{T}} \text{ and } \vec{\xi} = [\xi_1 \quad \xi_2 \quad \cdots \quad \xi_7]^{\mathrm{T}} \tag{5.26}$$

The notations concerning the weight matrices W_0 $(1\times 7$ matrix) and W $(7\times 3$ matrix) involved in the engine torque approximator are given below:

$$\mathrm{W}_0 = \begin{bmatrix} w_{01} & w_{02} & \cdots & w_{07} \end{bmatrix} \text{ and } \mathbb{W}_0 = \begin{bmatrix} w_{00} & \mathrm{W}_0 \end{bmatrix} \tag{5.27}$$

$$\begin{aligned} &\mathrm{W} = [\mathrm{W}_E \quad \mathrm{W}_{TC} \quad \mathrm{W}_R] \text{ and } \mathbb{W} = [\mathrm{W}_b \quad \mathrm{W}] \text{ where:} \\ &\mathrm{W}_E = \begin{bmatrix} w_{E,1} & w_{E,2} & \cdots & w_{E,7} \end{bmatrix}^{\mathrm{T}} \\ &\mathrm{W}_{TC} = \begin{bmatrix} w_{TC,1} & w_{TC,2} & \cdots & w_{TC,7} \end{bmatrix}^{\mathrm{T}} \\ &\mathrm{W}_R = \begin{bmatrix} w_{R,1} & w_{R,2} & \cdots & w_{R,7} \end{bmatrix}^{\mathrm{T}} \\ &\mathrm{W}_b = \begin{bmatrix} w_{b,1} & w_{b,2} & \cdots & w_{b,7} \end{bmatrix}^{\mathrm{T}} \end{aligned} \tag{5.28}$$

The notations concerning the weight matrices V_0 $(1\times 7$ matrix) and V $(7\times 3$ matrix) involved in the turbocharger torque approximator are given below:

$$\mathrm{V}_0 = \begin{bmatrix} v_{01} & v_{02} & \cdots & v_{07} \end{bmatrix} \text{ and } \mathbb{V}_0 = \begin{bmatrix} v_{00} & \mathrm{V}_0 \end{bmatrix} \tag{5.29}$$

$$\begin{aligned} &\mathrm{V} = [\mathrm{V}_E \quad \mathrm{V}_{TC} \quad \mathrm{V}_R] \text{ and } \mathbb{V} = [\mathrm{V}_b \quad \mathrm{V}] \text{ where:} \\ &\mathrm{V}_E = \begin{bmatrix} v_{E,1} & v_{E,2} & \cdots & v_{E,7} \end{bmatrix}^{\mathrm{T}} \\ &\mathrm{V}_{TC} = \begin{bmatrix} v_{TC,1} & v_{TC,2} & \cdots & v_{TC,7} \end{bmatrix}^{\mathrm{T}} \\ &\mathrm{V}_R = \begin{bmatrix} v_{R,1} & v_{R,2} & \cdots & v_{R,7} \end{bmatrix}^{\mathrm{T}} \\ &\mathrm{V}_b = \begin{bmatrix} v_{b,1} & v_{b,2} & \cdots & v_{b,7} \end{bmatrix}^{\mathrm{T}} \end{aligned} \tag{5.30}$$

Taking into account the notations introduced above, the neural torque approximators assume the following expressions as functions of the components of the input vector $\mathrm{X} = [N_E \quad N_{TC} \quad F_R]^{\mathrm{T}}$:

$$Q_E(N_E, N_{TC}, F_R) = \mathrm{Q_{Emax}} \cdot \{w_{00} + \sum_{i=1}^{7} w_{0i} \cdot \Phi(w_{b,i} + w_{E,i} \cdot N_E + w_{TC,i} \cdot N_{TC} + w_{R,i} \cdot F_R)\} \tag{5.31}$$

$$Q_{TC}(N_E, N_{TC}, F_R) = \mathrm{Q}_{\mathrm{Tmax}} \cdot \{v_{00} + \sum_{i=1}^{7} v_{0i} \cdot \Phi(v_{b,i} + v_{E,i} \cdot N_E + v_{TC,i} \cdot N_{TC} + v_{R,i} \cdot F_R)\} + \mathrm{Q}_{\mathrm{Cmax}} \tag{5.32}$$

The above torque approximator functions provide the required expressions for completing the state-space description of a typical marine plant. Additionally, functions $Q_E(\mathrm{X})$ and $Q_{TC}(\mathrm{X})$ are differentiable over the operating range of interest. Constant $\mathrm{Q}_{\mathrm{Emax}}$, used in the engine torque approximator, is required for assigning torque units to the neural net output, which is dimensionless. Moreover, as the net's output is normalised in the interval $\Im = [0,1]$, $\mathrm{Q}_{\mathrm{Emax}}$ can be set equal to the value of MCR engine torque delivery; practically, it is assigned a slightly larger value, so that the neural net is not operating in the vicinity of the saturation region of the sigmoid activation functions; this provision ensures enhanced numerical accuracy. In effect:

$$0 < Q_E(\mathrm{X}) \leq \mathrm{Q}_{\mathrm{Emax}}, \; \forall\, \mathrm{X} = [N \quad N_{TC} \quad F_R]^{\mathrm{T}} \tag{5.33}$$

In the same sense, constants $\mathrm{Q}_{\mathrm{Tmax}}$ and $\mathrm{Q}_{\mathrm{Cmax}}$, used in the turbocharger torque approximator, are employed for assigning torque units to the dimensionless neural net output. Additionally, the constants are employed in order to generate the actual values of turbocharger total torque Q_{TC}. In that respect, appropriate values for constants $\mathrm{Q}_{\mathrm{Tmax}}$ and $\mathrm{Q}_{\mathrm{Cmax}}$ are selected so that:

$$\mathrm{Q}_{\mathrm{Cmax}} \leq Q_{TC}(\mathrm{X}) \leq \mathrm{Q}_{\mathrm{Tmax}} + \mathrm{Q}_{\mathrm{Cmax}}, \; \forall\, \mathrm{X} = [N \quad N_{TC} \quad F_R]^{\mathrm{T}} \tag{5.34}$$

The selection can be done using the turbocharger torque map generated by the thermodynamic model, in order that the above inequality conditions are satisfied. Because values of turbocharger total torque, Q_{TC}, are both positive (turbo shaft acceleration) and negative (turbo shaft deceleration), and the neural net output lies in the interval $\Im = [0,1]$, it must hold that:

$$\mathrm{Q}_{\mathrm{Cmax}} < 0 \text{ and } \mathrm{Q}_{\mathrm{Tmax}} > |\mathrm{Q}_{\mathrm{Cmax}}| \tag{5.35}$$

Finally, as in the case of the engine torque approximator, appropriate selection of constants $\mathrm{Q}_{\mathrm{Tmax}}$ and $\mathrm{Q}_{\mathrm{Cmax}}$ can enhance numerical accuracy of the turbocharger torque approximator by preventing operation of the neural net's sigmoid activation functions in the saturation region.

In conclusion, the neural approximators can be used as substitutes of the engine and turbocharger torque maps. For each powerplant under consideration, however, the weight matrices $\mathbf{W}, \mathbf{W}_0$ and $\mathbf{V}, \mathbf{V}_0$ have to be known. The calibration of the neural approximators, also referred to by the term training, is, therefore, described in the next section.

5.2.2 Training of the Approximators

The determination of the weight matrices of any neural net can be performed in an automated manner by employing one of the training methods. In related literature, such as [48,49], a variety of training methods are described. There are two major categories of training algorithms: supervised and unsupervised training. Many authors also recognise a third category of training algorithms, referred to by the term self-supervised. In unsupervised training a sequence of input vectors is provided to the untrained net but no target outputs are specified. Then, the net is responsible of modifying the weights so that the most similar input vectors are assigned to the same output component or cluster unit. The above procedure indicates that nets where unsupervised training can be used are self-organising ones, in the sense that they group similar input vectors together without the use of training data to specify what a typical member of each group looks like or to which group each vector belongs.

In the case of the neural torque approximators, however, supervised training was used. This consists of the most typical neural net setting, and is performed by presenting a sequence of training vectors, or patterns, as inputs to the untrained net; each input vector is associated with a target output vector. The net's weights are adjusted according to a learning algorithm so that the net's output vector is as close as possible to the associated target output vector. The supervised training method with the widest acceptance is backpropagation. Indeed, in [48] it is reported that an estimate of 80% of all today's applications, utilising the backpropagation algorithm in one form or another, is quite realistic. Furthermore, and in spite of its limitations, backpropagation has dramatically expanded the range of problems to which neural computing methods can be applied, due to its sound mathematical background. In this respect, the Levenberg–Marquardt backpropagation training method has been the method of choice for the two neural torque approximators presented previously.

The weight matrices of the neural torque approximators $\underline{W}, \underline{W}_0$ and $\underline{V}, \underline{V}_0$ are determined by training the approximators using as training sets the torque maps generated by the numerical solution of the algebraic system of the thermodynamic engine model. As backpropagation was used, the weight final values were obtained by minimisation of the net's output error. Specifically, backpropagation forms a solution to the "weight-assignment-problem"; in fact, it is a mechanism to allocate the error observed at the output layer of the net between the two (or more) layers of weights. The mechanism calculates the value of weight modification so that the net's output Mean Square Error (MSE) in target function approximation is decreased after each epoch. An epoch is defined as a complete presentation of all the members of the training set to the neural net under training. The neural net training is dealt with as a minimisation problem, where the net MSE, at each epoch, is the objective function to be minimised; variables of this objective function are the weight values of the net synapses. Finally, a parameter called "learning rate", lying in the interval (0,1), is associated with backpropagation training. Learning rate determines how large the weight modifications can be. See [48,49] for more details on backpropagation and its variations.

The Levenberg–Marquardt backpropagation training method is named after the technique employed for the solution of the associated minimisation problem [15]. The technique is a compromise between Cauchy's gradient descent and Newton's techniques. It requires second-order information for the objective function (due to Newton's criterion evaluation), and calculates the direction of "descent" as a combination of Newton's and Cauchy's criteria. Since Newton's method is proven to perform better in the close vicinity of the optimum (minimum), whilst Cauchy's method works better during the initial steps of the iterative procedure, there is a constant which causes each one of the two methods combined to be enhanced or weakened according to whether the minimum is close or far. The Levenberg–Marquardt minimisation method has proven to work well with objective functions that are the sum of squared non-linear functions of the variables, such as the net's MSE. For a theoretical analysis see [15]. In [13] the Levenberg–Marquardt minimisation method is reported to be robust and iteratively more efficient; therefore it is indicated as the method of choice.

5.2.3 Typical Case Numerical Investigation

The training sets of the neural torque approximators for the test case of the MAN B&W 6L60MC propulsion plant were the engine and turbocharger torque maps generated by the numerical solution of the algebraic part of the thermodynamic model presented already in Chapter 2. The training grid of points was covering the range shown in Table 5.1. The same table also shows the MSE achieved and the learning rates used for training for the two approximators.

Table 5.1 Neural torque approximator training range and settings

	F_R (%)	N_E (rpm)	N_{TC} (rpm)
Min	30	65	2700
Max	100	120	14000
Step	5	5	500

	Eng. torq. approx.	T/C torq. approx.
Learn. rate	10^{-3}	10^{-3}
MSE achv'd	7.5×10^{-5}	3.0×10^{-6}

The plots in Figure 5.2 and Figure 5.3 provide neural approximator validation for operating points away from steady-state. Following the presentation strategy adopted in Chapter 2, the dependence of turbocharger total torque and engine torque delivery on turbocharger speed for six different settings of shaft speed and fuel index is depicted. The approximation error is also plotted. As can be seen, adequate accuracy is provided when using the approximators for torque calculation instead of the thermodynamic engine model. On the other hand, the approximator establishes an explicit, analytical mathematical dependence of torques on shaft/turbo rpm and index variables, which can be used in controller developments as explained in a later chapter.

In Table 5.2 the values of the weights and biases obtained after training are shown.

Table 5.2 Neural torque approximator weight and bias tables after training

Engine Torque Approximator: $Q_{Emax} = 900$ kN m, $w_{00} = 4.2723$					
i	W	W_E	W_{TC}	W_R	W_b
1	2.1319	0.0715	0.0037	5.1957	1.5798
2	0.1736	0.1296	0.0014	6.6596	8.7201
3	–4.2317	0.0167	0.0001	0.3045	–0.1855
4	–0.2969	0.1192	0.0018	14.3630	10.2347
5	–1.0880	0.0622	0.0010	8.9268	–8.0251
6	–7.0362	0.0098	0.0001	0.7085	0.4797
7	–0.2203	0.1664	0.0913	12.6202	–13.7209

Turbocharger Torque approximator: $Q_{Tmax} = 3000$ N m, $Q_{Cmax} = -2000$ N m, $v_{00} = 1.5599$					
i	V	V_E	V_{TC}	V_R	V_b
1	-3.2514	0.0272	0.0003	0.2592	–9.0349
2	-1.9519	–0.0233	–0.0001	–1.5399	4.5981
3	–0.0672	–0.0256	0.0008	–6.2438	0.9917
4	3.7274	0.0352	0.0002	1.4856	–9.8138
5	-3.7918	0.0218	–0.0000	1.4235	–4.9183
6	0.4951	0.0171	–0.0003	–1.1022	4.7128
7	0.6192	–0.0059	–0.0004	1.6420	4.2228

Finally, validation of the neural torque approximators against the thermodynamic model is given in Table 5.3 for steady-state.

Table 5.3 Steady-state validation of the neural torque approximators

Point	Engine torque Q_E (kN m)			Tubocharger torque Q_{TC} (N m)		
F_R (%)	Thermo	Neural	Error	Thermo	Neural	Error
60	450.600	450.0278	0.5722	3.5504	2.6491	0.9013
65	489.809	489.3354	0.4738	8.6139	7.3138	1.3001
70	529.061	528.6606	0.4009	10.5116	9.0547	1.4569
75	568.442	568.0799	0.3628	9.8707	9.2472	0.6235
80	607.738	607.3553	0.3826	11.3023	11.2984	0.0039
85	647.119	646.8605	0.2587	0.1537	0.0146	0.1391
90	686.586	686.2247	0.3616	4.6504	3.9371	0.7133
95	726.053	725.7865	0.2669	–6.1997	–7.6691	1.4694
100	765.520	764.9704	0.5502	6.2489	3.6524	2.5965

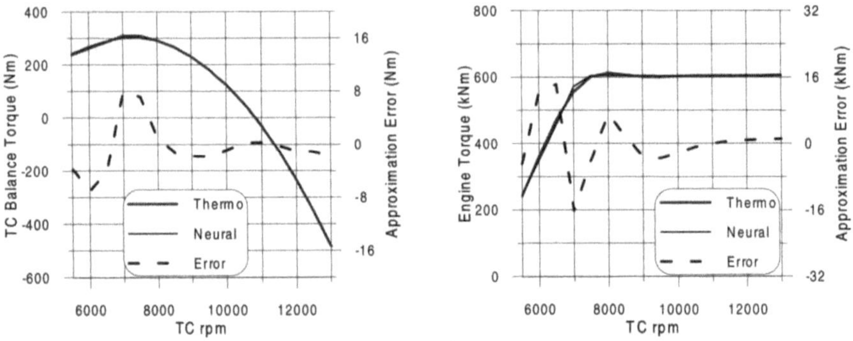

RPM = 110, %index = 80

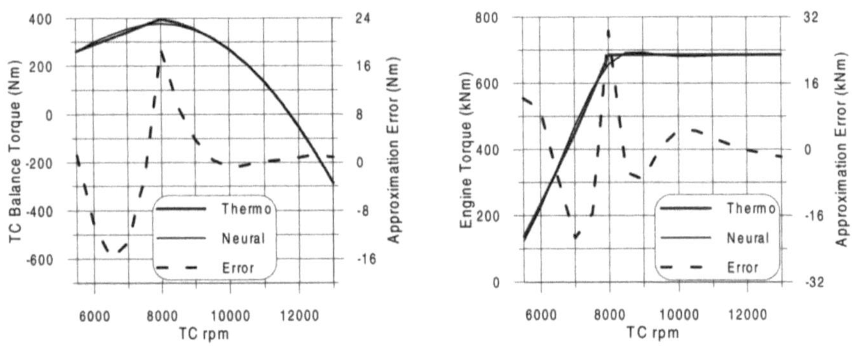

RPM = 110, %index = 90

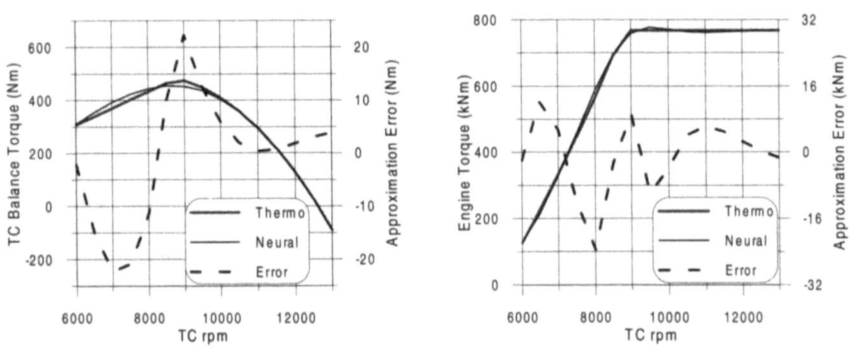

RPM = 110, %index = 100

Figure 5.2 Neural torque approximator validation away from steady-state

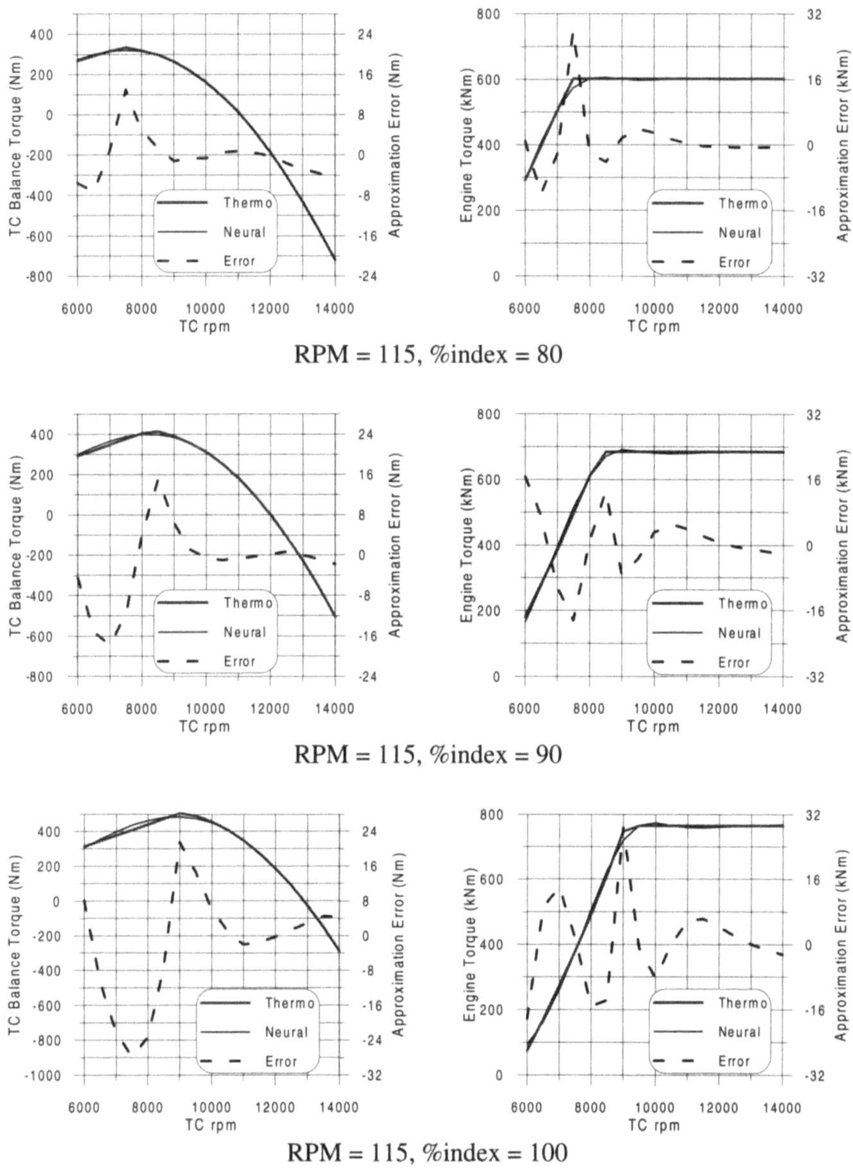

Figure 5.3 Neural torque approximator validation away from steady-state

5.3 State Equations of the Marine Plant

A general procedure for the configuration and training of the engine and turbocharger neural torque approximators based on maps obtained from physical models of Diesel ship propulsion plants has been presented in the previous section. The state-space equations were obtained in the form:

$$\dot{N} = \dot{N}_E = \frac{Q_E(N_E, N_{TC}, F_R) - Q_L(K_Q, N)}{I}$$
$$\dot{N}_{TC} = \frac{Q_{TC}(N_E, N_{TC}, F_R)}{I_{TC}} \tag{5.36}$$

where the functions $Q_E(N_E, N_{TC}, F_R)$ and $Q_{TC}(N_E, N_{TC}, F_R)$ are the neural torque approximators obtained previously. The propeller load torque $Q_L(N)$ can be calculated by the propeller law [29]:

$$Q_L(N) = K_Q \cdot N^2 = (K_{Q0} + k_Q) \cdot N^2 \tag{5.37}$$

where K_Q is the propeller-law constant. Note that although the propeller law is assumed to hold for steady-state operation it is extended to transient mode as well, in accordance with the quasi-steady modelling approach followed. However, to account for the disturbance introduced under the dynamic operating conditions of the propeller, the stochastic signal $k_Q(t)$ is superimposed on the nominal value K_{Q0}.

Engine–propeller shaft inertia I can be estimated by the following relation:

$$I = I_E + I_P = I_E + I_{P0} + \Delta I = I_0 + \Delta I \tag{5.38}$$

The nominal inertia component I_0 is calculated in the above equation as the sum of engine crankshaft and shafting system inertia I_E plus the nominal propeller inertia I_{P0}, which includes an amount accounting for the entrained water mass [29]. However, uncertainty ΔI is introduced in this parameter, as the entrained water mass varies significantly from the nominal value of $15\% I_{prop}$, where I_{prop} is the propeller "air" inertia. In [29] it is argued that uncertainty ΔI lies within the range $6-30\% I_{prop}$. Note that in the formulation of the dynamical equations of the engine thermodynamic model of Chapter 2 (group DYN) the uncertainty ΔI involved in the engine–propeller shaft inertia due to the entrained water mass has not been taken into account.

In conclusion, the state-space equations of the marine plant are as follows in matrix form:

$$\begin{bmatrix} \dot{N} \\ \dot{N}_{TC} \end{bmatrix} = \begin{bmatrix} \dfrac{\mathrm{Q}_{\mathrm{Emax}} \cdot \{w_{00} + \sum_{i=1}^{7} w_{0i} \cdot \Phi(w_{b,i} + w_{E,i} \cdot N + w_{TC,i} \cdot N_{TC} + w_{R,i} \cdot F_R)\} - N^2 \cdot K_Q}{I} \\ \dfrac{\mathrm{Q}_{\mathrm{Tmax}} \cdot \{v_{00} + \sum_{i=1}^{7} v_{0i} \cdot \Phi(v_{b,i} + v_{E,i} \cdot N + v_{TC,i} \cdot N_{TC} + v_{R,i} \cdot F_R)\} + \mathrm{Q}_{\mathrm{Cmax}}}{I_{TC}} \end{bmatrix} \tag{5.39}$$

The correspondence with the generic form of the state equations presented in Section 5.1.2 is completed if:

- The dynamical equation includes an l-dimensional vector ρ, "packing" plant uncertain parameters, as follows:

  $$\dot{\mathrm{x}} = \mathrm{f}(\mathrm{x},\mathrm{u},\rho) \tag{5.40}$$

  In the case of marine plants, the major sources of uncertainty are perturbation of propeller torque coefficient K_Q, and fluctuation of shafting system inertia I. Therefore:

  $$\rho^{\mathrm{T}} = [\, K_Q \quad I \,] \tag{5.41}$$

- The output vector y coincides with the state vector, *i.e.*:

  $$\mathrm{y}(t) \equiv \mathrm{x}(t) \Leftrightarrow \mathrm{h}(\mathrm{x}) = \begin{bmatrix} 1 & 0 \\ 0 & 1 \end{bmatrix} \tag{5.42}$$

  This means that the effect of the measuring devices is ignored. Although this may not always be the case, the following reasons support this modelling assumption:

  (i) Sensor dynamics have also been ignored during the construction of the state vector, as the only variables included are the ones dictated by the physical model for the process.

  (ii) It is aimed to keep this analysis simple and focused mainly on the control problems inherent in the marine plant. Extension of the methods presented here in order to include the sensor dynamics and non-linear features is the subject of future work.

In Section 5.4, the state-space equations presented here are linearised and decomposed, in order for the feedback (closed-loop) and feedforward (open-loop) control problems to be formulated.

5.4 State-space Decomposition and Uncertainty

5.4.1 Manipulation of Equations and Variables

Using the standard linearisation practice for the non-linear state equations that was presented previously and employed often for control purposes the plant signals, state variables and control actions are decomposed to a steady-state value and a superimposed perturbation signal. In this respect the following analysis of the shaft/turbo rpm and fuel index signals is done:

$$N = N_0 + n,\ N_{TC} = N_{TC0} + n_{TC},\ F_R = F_{R0} + f_R \tag{5.43}$$

where the subscript 0 denotes the steady-state value and the lower-case letter the perturbation. Additionally, the propeller torque coefficient K_Q is considered as a time varying signal, consisting of a steady-state equilibrium value plus a stochastic fluctuation, *i.e.* as demonstrated in the previous section $K_Q(t) = K_{Q0} + k_Q(\mathrm{t})$. The steady-state values for shaft rpm N_0, turbo rpm N_{TC0}, fuel index F_{R0}, and propeller torque coefficient K_{Q0} correspond to the equilibrium points of the plant. These points can be determined using the following conditions:

$$\left.\begin{array}{l}\dot{N} \equiv \dot{N}_E = 0\\ \dot{N}_{TC} = 0\end{array}\right\} \Leftrightarrow \left\{\begin{array}{l}Q_E(N_{E0}, N_{TC0}, F_{R0}) = Q_L(K_{Q0}, N_0)\\ Q_{TC}(N_{E0}, N_{TC0}, F_{R0}) = 0\end{array}\right. \tag{5.44}$$

Note that calculation of the equilibrium points is done from the marine plant equations without including any uncertainty, *i.e.* the *l*-dimensional vector ρ of system uncertain parameters is assumed to hold its nominal value $\rho_0^{\mathrm{T}} = [K_{Q0}\ \ I_0]$.

In the case of marine plants, it proves algebraically convenient for linearisation to consider the uncertainty in K_Q as plant disturbance. Therefore, a plant disturbance vector can be defined as follows:

$$\mathrm{d}^{\mathrm{T}} = [\ k_Q(\mathrm{t})\ \ 0\ ] \tag{5.45}$$

In effect, the state equation is rewritten as:

$$\dot{\mathrm{x}} = \mathrm{f}(\mathrm{x},\mathrm{u},\mathrm{d}) \tag{5.46}$$

Note that, although uncertainty ΔI is omitted, it is not dropped permanently, but just for the sake of simplifying the formulas.

Under this assumption, the previously presented generic linearisation equation, which was derived using a first-order Taylor expansion for f, is modified as follows in order to include the disturbance vector:

$$\mathrm{f}(\mathrm{x},\mathrm{u},\mathrm{d}) \approx \ddot{\mathrm{f}}'_{\mathrm{x}}(\mathrm{x}_0,\mathrm{u}_0,\vec{0})\cdot\delta\mathrm{x} + \ddot{\mathrm{f}}'_{\mathrm{u}}(\mathrm{x}_0,\mathrm{u}_0,\vec{0})\cdot\delta\mathrm{u} + \ddot{\mathrm{f}}'_{\mathrm{d}}(\mathrm{x}_0,\mathrm{u}_0,\vec{0})\cdot\mathrm{d} \tag{5.47}$$

where obviously $\delta\mathrm{d} \equiv \mathrm{d}$. The $(n \times n)$ derivative matrix $\ddot{\mathrm{f}}'_{\mathrm{d}}(\mathrm{x}_0,\mathrm{u}_0,\vec{0})$ is defined in analogy with the other two similar matrices, $\ddot{\mathrm{f}}'_{\mathrm{x}}$ and $\ddot{\mathrm{f}}'_{\mathrm{u}}$, as follows:

$$\ddot{\mathrm{f}}'_{\mathrm{d}}(\mathrm{x}_0,\mathrm{u}_0,\vec{0}) \triangleq \left[\left.\frac{\partial \mathrm{f}}{\partial d_1}\right|_{(\mathrm{x}_0,\mathrm{u}_0,\vec{0})} \quad \left.\frac{\partial \mathrm{f}}{\partial d_2}\right|_{(\mathrm{x}_0,\mathrm{u}_0,\vec{0})} \quad \cdots \quad \left.\frac{\partial \mathrm{f}}{\partial d_n}\right|_{(\mathrm{x}_0,\mathrm{u}_0,\vec{0})}\right] \tag{5.48}$$

In the case of the marine plant, the non-linear state-space equations are manipulated in order to provide a linearised perturbation part, in which only the perturbation signals are involved. Therefore:

$$\begin{bmatrix} I\cdot\dot{N} \\ I_{TC}\cdot\dot{N}_{TC}\end{bmatrix}=$$

$$\begin{bmatrix} \left.\frac{\partial Q_E}{\partial N}\right|_{X_0}-\left.\frac{\partial Q_L}{\partial N}\right|_{N_0,K_{Q0}} & \left.\frac{\partial Q_E}{\partial N_{TC}}\right|_{X_0} \\ \left.\frac{\partial Q_{TC}}{\partial N_E}\right|_{X_0} & \left.\frac{\partial Q_{TC}}{\partial N_{TC}}\right|_{X_0}\end{bmatrix}\cdot\begin{bmatrix} n \\ n_{TC}\end{bmatrix}+\begin{bmatrix}\left.\frac{\partial Q_E}{\partial F_R}\right|_{X_0} \\ \left.\frac{\partial Q_{TC}}{\partial F_R}\right|_{X_0}\end{bmatrix}\cdot f_R+\begin{bmatrix}-\left.\frac{\partial Q_L}{\partial K_Q}\right|_{N_0,K_{Q0}} & 0 \\ 0 & 0\end{bmatrix}\cdot\begin{bmatrix} k_Q \\ 0\end{bmatrix} \tag{5.49}$$

Derivation of the above is explained hereafter, along with the calculation of the partial derivatives involved. In order to formulate the linearisation equation from the plant state-space description, the torque variables (engine, turbocharger and propeller) are expanded to Taylor series and the first-order term is kept. Specifically, for the engine and turbocharger torques:

$$\begin{aligned} Q_E &\approx Q_{E0}+\begin{bmatrix} q_{E,1} & q_{E,2}\end{bmatrix}\cdot\begin{bmatrix} n \\ n_{TC}\end{bmatrix}+q_{E,0}\cdot f_R \\ Q_{TC} &\approx Q_{TC0}+\begin{bmatrix} q_{TC,1} & q_{TC,2}\end{bmatrix}\cdot\begin{bmatrix} n \\ n_{TC}\end{bmatrix}+q_{TC,0}\cdot f_R \end{aligned} \tag{5.50}$$

where $q_{E,0}, q_{E,1}, q_{E,2}, q_{TC,0}, q_{TC,1}, q_{TC,2}$ are the partial derivatives of the torque approximator functions with respect to the variable of interest (shaft/turbo rpm, fuel index) calculated at an equilibrium point determined by the vector $X_0=[N_0 \quad N_{TC0} \quad F_{R0}]^T$.

Specifically, the partial derivatives of engine torque are calculated as follows:

$$q_{E,0}=\left.\frac{\partial}{\partial F_R}Q_E(N,N_{TC},F_R)\right|_{X=X_0}=Q_{Emax}\cdot\left\{W_0\cdot\left(\frac{\partial}{\partial F_R}\Phi(W_b+W\cdot X)\right)_{X=X_0}\right\} \tag{5.51}$$

$$q_{E,1}=\left.\frac{\partial}{\partial N}Q_E(N,N_{TC},F_R)\right|_{X=X_0}=Q_{Emax}\cdot\left\{W_0\cdot\left(\frac{\partial}{\partial N}\Phi(W_b+W\cdot X)\right)_{X=X_0}\right\} \tag{5.52}$$

$$q_{E,2}=\left.\frac{\partial}{\partial N_{TC}}Q_E(N,N_{TC},F_R)\right|_{X=X_0}=Q_{Emax}\cdot\left\{W_0\cdot\left(\frac{\partial}{\partial N_{TC}}\Phi(W_b+W\cdot X)\right)_{X=X_0}\right\} \tag{5.53}$$

The partial derivatives of turbocharger torque are computed as follows:

$$q_{TC,0}=\left.\frac{\partial}{\partial F_R}Q_{TC}(N,N_{TC},F_R)\right|_{X=X_0}=Q_{Tmax}\cdot\left\{V_0\cdot\left(\frac{\partial}{\partial F_R}\Phi(V_b+V\cdot X)\right)_{X=X_0}\right\} \tag{5.54}$$

$$q_{TC,1}=\left.\frac{\partial}{\partial N}Q_{TC}(N,N_{TC},F_R)\right|_{X=X_0}=Q_{Tmax}\cdot\left\{V_0\cdot\left(\frac{\partial}{\partial N}\Phi(V_b+V\cdot X)\right)_{X=X_0}\right\} \tag{5.55}$$

$$q_{TC,2}=\left.\frac{\partial}{\partial N_{TC}}Q_{TC}(N,N_{TC},F_R)\right|_{X=X_0}=Q_{Tmax}\cdot\left\{V_0\cdot\left(\frac{\partial}{\partial N_{TC}}\Phi(V_b+V\cdot X)\right)_{X=X_0}\right\} \tag{5.56}$$

The partial derivatives of the vector field $\Phi(\vec{\xi})$, appearing in the above equations, are computed according to:

$$\frac{\partial}{\partial X_j}\Phi(\mathrm{L}_0+\mathrm{L}\cdot\mathrm{X})=\begin{bmatrix} L_{1j}\cdot\Phi'(L_{0,1}+\begin{bmatrix}L_{11} & L_{12} & L_{13}\end{bmatrix}\cdot\mathrm{X}) \\ L_{2j}\cdot\Phi'(L_{0,2}+\begin{bmatrix}L_{21} & L_{22} & L_{23}\end{bmatrix}\cdot\mathrm{X}) \\ \vdots \\ L_{7j}\cdot\Phi'(L_{0,7}+\begin{bmatrix}L_{71} & L_{72} & L_{73}\end{bmatrix}\cdot\mathrm{X}) \end{bmatrix}$$

where (5.57)

$$\mathrm{L}=\begin{bmatrix} L_{11} & L_{12} & L_{13} \\ L_{21} & L_{22} & L_{23} \\ & \vdots & \\ L_{71} & L_{72} & L_{73} \end{bmatrix} \text{ and } \mathrm{L}_0=\begin{bmatrix} L_{0,1} \\ L_{0,2} \\ \vdots \\ L_{0,7} \end{bmatrix}$$

Use of the following fact of differential calculus for computing the derivative of the real-valued logistic sigmoid function is employed:

$$\Phi'(\xi)=\frac{d}{d\xi}\Phi(\xi)=\frac{d}{d\xi}\left(\frac{1}{1+e^{-\xi}}\right)=e^{-\xi}\cdot\Phi^2(\xi)=\frac{\Phi^2(\xi)}{\exp(\xi)} \tag{5.58}$$

In conclusion, the following analytical expressions are obtained for the six torque derivative parameters $q_{E,0}, q_{E,1}, q_{E,2}, q_{TC,0}, q_{TC,1}, q_{TC,2}$ appearing in the Taylor series expansion of the two torque approximator functions:

$$q_{E,0}=\mathrm{Q}_{\mathrm{Emax}}\cdot\sum_{i=1}^{7}w_{0i}w_{R,i}\cdot\frac{\Phi^2(w_{E,i}\cdot N_0+w_{TC,i}\cdot N_{TC0}+w_{R,i}\cdot F_{R0}+w_{b,i})}{\exp(w_{E,i}\cdot N_0+w_{TC,i}\cdot N_{TC0}+w_{R,i}\cdot F_{R0}+w_{b,i})} \tag{5.59}$$

$$q_{E,1}=\mathrm{Q}_{\mathrm{Emax}}\cdot\sum_{i=1}^{7}w_{0i}w_{E,i}\cdot\frac{\Phi^2(w_{E,i}\cdot N_0+w_{TC,i}\cdot N_{TC0}+w_{R,i}\cdot F_{R0}+w_{b,i})}{\exp(w_{E,i}\cdot N_0+w_{TC,i}\cdot N_{TC0}+w_{R,i}\cdot F_{R0}+w_{b,i})} \tag{5.60}$$

$$q_{E,2}=\mathrm{Q}_{\mathrm{Emax}}\cdot\sum_{i=1}^{7}w_{0i}w_{TC,i}\cdot\frac{\Phi^2(w_{E,i}\cdot N_0+w_{TC,i}\cdot N_{TC0}+w_{R,i}\cdot F_{R0}+w_{b,i})}{\exp(w_{E,i}\cdot N_0+w_{TC,i}\cdot N_{TC0}+w_{R,i}\cdot F_{R0}+w_{b,i})} \tag{5.61}$$

$$q_{TC,0}=\mathrm{Q}_{\mathrm{Tmax}}\cdot\sum_{i=1}^{7}v_{0i}v_{R,i}\cdot\frac{\Phi^2(v_{E,i}\cdot N_0+v_{TC,i}\cdot N_{TC0}+v_{R,i}\cdot F_{R0}+v_{b,i})}{\exp(v_{E,i}\cdot N_0+v_{TC,i}\cdot N_{TC0}+v_{R,i}\cdot F_{R0}+v_{b,i})} \tag{5.62}$$

$$q_{TC,1}=\mathrm{Q}_{\mathrm{Tmax}}\cdot\sum_{i=1}^{7}v_{0i}v_{E,i}\cdot\frac{\Phi^2(v_{E,i}\cdot N_0+v_{TC,i}\cdot N_{TC0}+v_{R,i}\cdot F_{R0}+v_{b,i})}{\exp(v_{E,i}\cdot N_0+v_{TC,i}\cdot N_{TC0}+v_{R,i}\cdot F_{R0}+v_{b,i})} \tag{5.63}$$

$$q_{TC,2}=\mathrm{Q}_{\mathrm{Tmax}}\cdot\sum_{i=1}^{7}v_{0i}v_{TC,i}\cdot\frac{\Phi^2(v_{E,i}\cdot N_0+v_{TC,i}\cdot N_{TC0}+v_{R,i}\cdot F_{R0}+v_{b,i})}{\exp(v_{E,i}\cdot N_0+v_{TC,i}\cdot N_{TC0}+v_{R,i}\cdot F_{R0}+v_{b,i})} \tag{5.64}$$

Finally, the decomposition in the case of propeller load torque $Q_L(K_Q,N)$ is as follows:

$$Q_L=K_Q\cdot N^2\approx K_{Q0}\cdot N_0^2+N_0^2\cdot k_Q+2K_{Q0}N_0\cdot n=Q_{L0}+N_0^2\cdot k_Q+2K_{Q0}N_0\cdot n \tag{5.65}$$

where obviously $Q_{L0}=Q_L(K_{Q0},N_0)=K_{Q0}\cdot N_0^2$ is the nominal, steady-state propeller load torque, which the powerplant has been designed to accept. Note that the term proportional to the quantity $(k_Q \cdot n)$ has not been included in the above expression, as it is a second-order term in the Taylor expansion of Q_L.

The steady-state values Q_{E0}, Q_{L0} and Q_{TC0} can be determined using the equilibrium conditions at steady-state, as already pointed out; it holds that:

$$Q_{E0} = Q_E(N_{E0}, N_{TC0}, F_{R0}) = Q_{L0} = Q_L(K_{Q0}, N_0) = K_{Q0} \cdot N_0^2 \tag{5.66}$$

and

$$Q_{TC0} = Q_{TC}(N_{E0}, N_{TC0}, F_{R0}) = 0 \tag{5.67}$$

In effect, the linearisation equation takes the following form:

$$\begin{bmatrix} I \cdot \dot{n} \\ I_{TC} \cdot \dot{n}_{TC} \end{bmatrix} = \begin{bmatrix} q_{E,1} - 2K_{Q0}N_0 & q_{E,2} \\ q_{TC,1} & q_{TC,2} \end{bmatrix} \cdot \begin{bmatrix} n \\ n_{TC} \end{bmatrix} + \begin{bmatrix} q_{E,0} \\ q_{TC,0} \end{bmatrix} \cdot f_R + \begin{bmatrix} -N_0^2 \cdot k_Q \\ 0 \end{bmatrix} \tag{5.68}$$

Note that in the above derivatives $\dot{N}$ and $\dot{N}_{TC}$ have been replaced by $\dot{n}$ and $\dot{n}_{TC}$ respectively. This is due to the fact that for the operating point regulation problem of the marine plant, which is addressed extensively in the rest of this text, the steady-state operating point is assumed not to change, *i.e.*:

$$\dot{N}_0 \equiv \dot{N}_{E0} = 0 \text{ and } \dot{N}_{TC0} = 0 \tag{5.69}$$

Furthermore, note that the expression:

$$\vec{f}_d'(x_0, u_0, \vec{0}) \cdot d = \begin{bmatrix} -\dfrac{\partial Q_L}{\partial K_Q}\bigg|_{N_0, K_{Q0}} & 0 \\ 0 & 0 \end{bmatrix} \cdot \begin{bmatrix} k_Q \\ 0 \end{bmatrix}$$

has been substituted by the vector $d^T = \begin{bmatrix} -N_0^2 \cdot k_Q & 0 \end{bmatrix}$ because the major disturbance component in the case of marine plants appears only in the first dynamical equation, *i.e.* the one concerning shaft rpm. Additionally, the order of the state vector is low (equal to two). Therefore, for the sake of simplicity, it is better to "pack" the derivative matrix $\vec{f}_d'(x_0, u_0, \vec{0})$ with the disturbance vector d. As will be seen later, this also facilitates formulation of the transfer function matrix and controller design.

5.4.2 State-space Parametric Uncertainty and Disturbance

The previously deduced linearised equation holds when the operating point is not assumed to change, *i.e.* for the speed regulation problem. However, marine plants do experience rather slow changes in terms of operating point, when engine loading or unloading is required. Therefore, the generalisation of the above linearisation methodology is attempted here. Furthermore, the inherent uncertainty of the typical marine plant is identified and investigated.

The initial non-linear state equations:

$$\dot{N} = \dot{N}_E = \frac{Q_E(N_E, N_{TC}, F_R) - Q_L(K_Q, N)}{I}$$
$$\dot{N}_{TC} = \frac{Q_{TC}(N_E, N_{TC}, F_R)}{I_{TC}} \tag{5.70}$$

are now decomposed to a "nominal" and a "perturbation" part, without the assumptions that $\dot{N}_0 \equiv \dot{N}_{E0} = 0$ and $\dot{N}_{TC0} = 0$, as in the case of the regulation problem. In effect, we obtain that:

$$(I_0 + \Delta I) \cdot (\dot{N}_0 + \dot{n}) = Q_E(N_0, N_{TC0}, F_{R0}) - Q_L(K_{Q0}, N_0)$$
$$+ \begin{bmatrix} q_{E,1} - 2K_{Q0}N_0 & q_{E,2} \end{bmatrix} \cdot \begin{bmatrix} n \\ n_{TC} \end{bmatrix} + q_{E,0} \cdot f_R - N_0^2 \cdot k_Q \tag{5.71}$$
$$I_{TC} \cdot (\dot{N}_{TC0} + \dot{n}_{TC}) = Q_{TC}(N_0, N_{TC0}, F_{R0}) + \begin{bmatrix} q_{TC,1} & q_{TC,2} \end{bmatrix} \cdot \begin{bmatrix} n \\ n_{TC} \end{bmatrix} + q_{TC,0} \cdot f_R$$

Based on the complete non-linear equations above and the signal decomposition scheme given earlier, the two dynamical state equations of the marine propulsion plant are separated to form the following two independent models:

(a) Nominal Non-linear Model (N^2M):

$$\dot{N}_0 = \frac{Q_E(N_0, N_{TC0}, F_{R0}) - K_{Q0} \cdot N_0^2}{I_0}$$
$$\dot{N}_{TC0} = \frac{Q_{TC}(N_0, N_{TC0}, F_{R0})}{I_{TC}} \tag{5.72}$$

The torque values $Q_E(N_0, N_{TC0}, F_{R0})$ and $Q_{TC}(N_0, N_{TC0}, F_{R0})$ are calculated as functions of the shaft/turbo rpm and fuel index values, propagated through the neural torque approximators, *i.e.*:

$$\frac{Q_E(N_0, N_{TC0}, F_{R0})}{\mathrm{Q_{Emax}}} = w_{00} + \sum_{i=1}^{7} w_{0i} \cdot \Phi(w_{E,i} \cdot N_0 + w_{TC,i} \cdot N_{TC0} + w_{R,i} \cdot F_{R0} + w_{b,i}) \tag{5.73}$$

$$\frac{Q_{TC}(N_0, N_{TC0}, F_{R0}) - \mathrm{Q_{Cmax}}}{\mathrm{Q_{Tmax}}} = v_{00} + \sum_{i=1}^{7} v_{0i} \cdot \Phi(v_{E,i} \cdot N_0 + v_{TC,i} \cdot N_{TC0} + v_{R,i} \cdot F_{R0} + v_{b,i}) \tag{5.74}$$

Note that the values calculated now generally differ from the steady-state values Q_{E0}, Q_{L0} and Q_{TC0} presented previously, because the latter are calculated only at steady-state points, at which the equilibrium condition of zero acceleration is assumed to hold, *i.e.* $\dot{N}_0 = 0$ and $\dot{N}_{TC0} = 0$. On the other hand, in this case, the torque values are determined by the input vector $\mathrm{X}_0 = [N_0 \quad N_{TC0} \quad F_{R0}]^{\mathrm{T}}$. For example open-loop fuel index schedules $F_{R0}(t)$ can be applied when a relatively slow and gradual change of the operating point is required. In that case, the rates of shaft and turbo rpm change, $\dot{N}_0$ and $\dot{N}_{TC0}$, are determined by the equations of the N^2M, which incorporate the neural torque approximators, the nominal propeller law and nominal shaft inertia. Therefore, the system description is rather certain; results from supervisory or non-linear control theory can be applied. Such control options of the N^2M are examined in upcoming chapters.

(b) Uncertain Perturbation Model (UPM):

$$\begin{aligned} \dot{N}_0 \cdot \Delta I + I \cdot \dot{n} &= \begin{bmatrix} q_{E,1} - 2K_{Q0}N_0 & q_{E,2} \end{bmatrix} \cdot \begin{bmatrix} n \\ n_{TC} \end{bmatrix} + q_{E,0} \cdot f_R - N_0^2 \cdot k_Q \\ I_{TC} \cdot \dot{n}_{TC} &= \begin{bmatrix} q_{TC,1} & q_{TC,2} \end{bmatrix} \cdot \begin{bmatrix} n \\ n_{TC} \end{bmatrix} + q_{TC,0} \cdot f_R \end{aligned} \tag{5.75}$$

In matrix form the above perturbation equations can be written as follows:

$$\begin{bmatrix} \dot{n} \\ \dot{n}_{TC} \end{bmatrix} = \begin{bmatrix} \dfrac{q_{E,1} - 2K_{Q0}N_0}{I} & \dfrac{q_{E,2}}{I} \\ \dfrac{q_{TC,1}}{I_{TC}} & \dfrac{q_{TC,2}}{I_{TC}} \end{bmatrix} \cdot \begin{bmatrix} n \\ n_{TC} \end{bmatrix} + \begin{bmatrix} \dfrac{q_{E,0}}{I} \\ \dfrac{q_{TC,0}}{I_{TC}} \end{bmatrix} \cdot f_R + \begin{bmatrix} -\dfrac{N_0^2 \cdot k_Q + \dot{N}_0 \cdot \Delta I}{I} \\ 0 \end{bmatrix} \tag{5.76}$$

or more compactly:

$$\delta\dot{\mathrm{x}}(t) = \mathrm{A} \cdot \delta\mathrm{x}(t) + \mathrm{B} \cdot \delta u(t) + \mathrm{d}(t)$$

This is the dynamic state (vector) equation of a Single-Input, Multi-Output (SIMO, (1×2)) linear system with scalar disturbance [50]. The symbol correspondence is straightforward and therefore omitted. Furthermore, as analysed hereafter, the above linear perturbation system is uncertain.

In addition to the inherent uncertainty in the shaft total inertia I, due to the variation of the entrained water mass, there are other sources of parametric uncertainty. First, the values of the partial derivatives of the torque approximators employed for linearisation are varying; their value, as already analysed, depends upon the input vector value $\mathrm{X}_0 = [N_0 \quad N_{TC0} \quad F_{R0}]^{\mathrm{T}}$. In order to obtain a global model valid over the complete operating range of the plant, the following uncertainty is introduced in the partial derivative coefficients:

$$q_{E,i} = q_{E0,i} + \Delta q_{E,i},\; q_{TC,i} = q_{TC0,i} + \Delta q_{TC,i} \text{ for } i = 0,1,2 \tag{5.77}$$

Moreover, the (1,1)-element of transient matrix A (a_{11}) carries additional uncertainty due to the variance of $\partial Q_L / \partial N = 2K_{Q0}N_0$.

Note that even if partial derivatives $q_{E,i}, q_{TC,i}$, $i = 0,1,2$ and/or $\partial Q_L / \partial N$ did not exhibit significant fluctuation, from one operating point to another, the quasi-steady, cycle-mean thermodynamic modelling approach used involves some uncertainty due to the modelling assumptions adopted. Furthermore, the use of neural nets in order to approximate the torque functions, which are mathematically unknown (only the points on the torque maps can be considered known), introduces further error. Therefore, the introduction of the uncertainties above has the positive side effect of incorporating, in some degree, the modelling uncertainty of the quasi-steady assumption, as well as the approximation error due to the use of the neural torque approximators [50].

A systematic manner of calculation of the uncertainty bounds for the elements of the perturbation system matrices A and B can be effectuated using the mathematical formulation presented hereafter. As an indicative example, the case of the (1,1)-element of transient matrix A (a_{11}) is given. As already argued, a_{11} is a real-valued function defined on subspace U of $\mathbb{R}^3$ as follows:

$$a_{11}: \ \mathrm{U} \subset \mathbb{R}^3 \rightarrow \mathbb{R}$$

$$\begin{bmatrix} q_{E,1} \\ N_0 \\ I \end{bmatrix} \mapsto a_{11}(q_{E,1}, N_0, I) = \frac{q_{E,1} - 2K_{Q0}N_0}{I} \tag{5.78}$$

Subspace U is defined as follows:

$$\mathrm{U} = \{\vec{\xi} \in \mathbb{R}^3 : \ q_{E0,1} - \delta q_{E,1} \le \xi_1 \le q_{E0,1} + \delta q_{E,1}, \quad N_{00} - \delta N_0 \le \xi_2 \le N_{00} + \delta N_0, \ I_0 - \delta I \le \xi_3 \le I_0 + \delta I\} \tag{5.79}$$

where lower case delta denotes the uncertainty “radius” of each variable as defined hereafter:

$$\delta q_{E,1} = \max |\Delta q_{E,1}|, \ \delta N_0 = \max |\Delta N_0|, \ \delta I = \max |\Delta I| \tag{5.80}$$

Shaft rpm “uncertainty” ΔN_0 is introduced so that the derived uncertain linear perturbation model is valid over the entire operating range of interest and not localised around a certain operating point. In the same sense, a “nominal” value of shaft rpm N_{00} can be selected at the centre of the operating region, which the model intends to describe, or for which the feedback controller aims to perform disturbance rejection.

Uncertainty in the engine torque approximator partial derivative $\Delta q_{E,1}$, and in effect bound $\delta q_{E,1}$, as well as a nominal value for the derivative itself $q_{E0,1}$, can be calculated using the neural engine torque approximator $Q_E(N, N_{TC}, F_R)$. Furthermore, the calculation can be done either analytically or numerically. Owing to the availability of computers with enhanced capabilities the numerical approach is preferable. Specifically, a map can be constructed for a fine grid of points, defined for various values of the input vector $\mathrm{X} = [N \ \ N_{TC} \ \ F_R]^{\mathrm{T}}$; at each point the value of $q_{E,1}$ is calculated by the relationship given before:

$$q_{E,1} = \frac{\partial}{\partial N} Q_E(N, N_{TC}, F_R) =$$
$$\mathrm{Q_{Emax}} \cdot \sum_{i=1}^{7} w_{0i} w_{E,i} \cdot \frac{\Phi^2(w_{E,i} \cdot N_0 + w_{TC,i} \cdot N_{TC0} + w_{R,i} \cdot F_{R0} + w_{b,i})}{\exp(w_{E,i} \cdot N_0 + w_{TC,i} \cdot N_{TC0} + w_{R,i} \cdot F_{R0} + w_{b,i})} \tag{5.81}$$

Then, the maximum and minimum values, $q_{E\max,1}$ and $q_{E\min,1}$ respectively. can be determined. In effect, it can be set that:

$$\delta q_{E,1} = \frac{q_{E\max,1} - q_{E\min,1}}{2} \text{ and } q_{E0,1} = q_{E\min,1} + \delta q_{E,1} \tag{5.82}$$

Note that the above method can be employed because the torque approximator functions are differentiable and the mathematical expressions of their partial derivatives allow explicit calculation at any point of interest. Due to continuity arguments, the finer the grid of points used, the more accurate is the determination of the maximum and minimum values, $q_{E\max,1}$ and $q_{E\min,1}$. To begin with, the grid to use can be the same one used for the iterative numerical solution of the algebraic equations of the engine thermodynamic model and the generation of the training sets of the two approximators.

Alternatively, the extremum values of function $q_{E,1}(N, N_{TC}, F_R)$, for the operating range of interest, can be determined using analytical mathematical techniques. As can be seen, function $q_{E,1}(N, N_{TC}, F_R)$ is continuous and differentiable over the plant's operating range. Therefore, the method of zero gradient can be applied. To be specific, local maxima (minima) can be determined using the conditions:

- $\frac{\partial}{\partial \mathrm{X}} q_{E,1}(N, N_{TC}, F_R) = \vec{0}$ and
- matrix $\left[\frac{\partial^2}{\partial \mathrm{X}_i \partial \mathrm{X}_j} q_{E,1}(N, N_{TC}, F_R) \right]$, $i, j = 1,2,3$ is negative (positive) definite.

Additionally, points at the bounds of the allowed operating region for $\mathrm{X} = [N \quad N_{TC} \quad F_R]^{\mathrm{T}}$ should be checked whether they are extrema or not. Then the global (in the region of interest) minimum and maximum of function $q_{E,1}(N, N_{TC}, F_R)$ can be sorted out. The analytical method, however, in spite its mathematical austerity, is rather tedious, as the calculation of partial derivatives of $q_{E,1}(N, N_{TC}, F_R)$ requires a significant amount of algebraic manipulation. Therefore, it will not be used or commented on any further.

In conclusion, the same procedure can be applied for estimating the uncertainty "radii" $\delta q_{E,i}$, $\delta q_{TC,i}$ and nominal values $q_{E0,i}$, $q_{TC0,i}$ for $i = 0,1,2$. In turn, this is required for calculating the uncertainty in the elements of propulsion systems matrices A and B, as in the case of a_{11}.

The physical bound for shaft inertia uncertainty ΔI, due to the effect of the entrained water mass, can be calculated, according to naval architecture literature, taking into account that the inertia added by the entrained water lies in the range $6-30\%I_{prop}$. By taking into consideration that the nominal inertia I_0 is already augmented by $15\%I_{prop}$, accounting for the "nominal" entrained water mass, a good choice for the uncertainty radius of inertia I is $\delta I = 15\%I_{prop}$. Note that the lower value for I is therefore equal to $I_0 - \delta I = I_E + I_{prop}$, corresponding to the case of zero water mass entrapment. Although, this is unrealistic in most normal sailing conditions, it can occur when the ship is sailing in heavy weather and rough seas; propeller ventilation can be partial or total, leading to a shaft inertia value lower than the value indicated above by the uncertainty bound.

After the quantification of uncertainties inherent in the physical parameters appearing in function $a_{11}(q_{E,1}, N_0, I)$, the uncertainty radius, δa_{11}, and nominal value, $a_{11,0}$ of this element of system's transient matrix, A, are calculated. With the rationale presented before, the radius and nominal value required can be determined if the minimum and maximum values, $a_{11,\min}$ and $a_{11,\max}$ respectively, can be obtained. Using the conditions of zero derivative, the following necessary condition is deduced for the local maxima or minima of the function:

$$\begin{bmatrix} \dfrac{\partial a_{11}}{\partial q_{E,1}} & \dfrac{\partial a_{11}}{\partial N_0} & \dfrac{\partial a_{11}}{\partial I} \end{bmatrix}^{\mathrm{T}} = \begin{bmatrix} \dfrac{1}{I} & -\dfrac{2K_{Q0}}{I} & -\dfrac{q_{E,1} - 2K_{Q0}N_0}{I^2} \end{bmatrix}^{\mathrm{T}} = \vec{0} \tag{5.83}$$

As can be seen from the above expression, however, the first and second components of the gradient vector above cannot be equal to zero in the operating range of interest. Therefore, the above necessary condition cannot hold anywhere in the interior of the operating range. Therefore, the maximum and minimum value of $a_{11}(q_{E,1}, N_0, I)$ has to be attained on the boundary of U (note that $a_{11}(q_{E,1}, N_0, I)$ is continuous and non-singular over entire U).

Due to the rpm-dependent propeller law, marine plants are stable, meaning that:

$$q_{E,1} < 2K_{Q0}N_0 \Rightarrow q_{E,1} - 2K_{Q0}N_0 < 0 \tag{5.84}$$

By taking into account the fact that total shaft inertia I is positive, the following bounds can be obtained for function $a_{11}(q_{E,1}, N_0, I)$:

$$a_{11,\min} = \frac{(q_{E0,1} - \delta q_{E,1}) - 2K_{Q0} \cdot (N_{00} + \delta N_0)}{I_0 - \delta I} \tag{5.85}$$

and

$$a_{11,\max} = \frac{(q_{E0,1} + \delta q_{E,1}) - 2K_{Q0} \cdot (N_{00} - \delta N_0)}{I_0 + \delta I} \tag{5.86}$$

Therefore:

$$\delta a_{11} = \frac{a_{11,\max} - a_{11,\min}}{2} \tag{5.87}$$

The same procedure can be applied for all the elements of system matrices A and B. In effect, the minimum and maximum values, $a_{ij,\min}$ and $a_{ij,\max}$, as well as $b_{i,\min}$ and $b_{i,\max}$, $i, j = 1,2$, are determined, and consequently the uncertainty radii δa_{ij} and δb_i, from the equation:

$$\delta a_{ij} = \frac{a_{ij,\max} - a_{ij,\min}}{2} \text{ or } \delta b_i = \frac{b_{i,\max} - b_{i,\min}}{2} \tag{5.88}$$

Taking into account the uncertainty radii of the physical parameters $q_{E,i}$, $q_{TC,i}$, $i = 0,1,2$, as well as total shaft inertia I, and shaft rpm N, the following ranges can be calculated for the system matrices:

$$\begin{aligned}
\mathrm{A}_{\min} &= \begin{bmatrix} \dfrac{(q_{E0,1} - \delta q_{E,1}) - 2K_{Q0}\cdot(N_{00} + \delta N_0)}{I_0 - \delta I} & \dfrac{q_{E0,2} - \delta q_{E,2}}{I_0 + \delta I} \\ \dfrac{q_{TC0,1} - \delta q_{TC,1}}{I_{TC}} & \dfrac{q_{TC0,2} - \delta q_{TC,2}}{I_{TC}} \end{bmatrix} \\
\mathrm{A}_{\max} &= \begin{bmatrix} \dfrac{(q_{E0,1} + \delta q_{E,1}) - 2K_{Q0}\cdot(N_{00} - \delta N_0)}{I_0 + \delta I} & \dfrac{q_{E0,2} + \delta q_{E,2}}{I_0 - \delta I} \\ \dfrac{q_{TC0,1} + \delta q_{TC,1}}{I_{TC}} & \dfrac{q_{TC0,2} + \delta q_{TC,2}}{I_{TC}} \end{bmatrix}
\end{aligned} \tag{5.89}$$

and

$$\mathrm{B}_{\min} = \begin{bmatrix} \dfrac{q_{E0,0} - \delta q_{E,0}}{I_0 - \delta I} \\ \dfrac{q_{TC,0} - \delta q_{TC,0}}{I_{TC}} \end{bmatrix}, \quad \mathrm{B}_{\max} = \begin{bmatrix} \dfrac{q_{E0,0} + \delta q_{E,0}}{I_0 - \delta I} \\ \dfrac{q_{TC0,0} + \delta q_{TC,0}}{I_{TC}} \end{bmatrix} \tag{5.90}$$

For the above bounds to hold, the following assumptions have been made, which are valid for typical marine propulsion powerplants and engines:

- $0 \le q_{E,1} < 2K_{Q0}N_0$: as already explained, this is due to the inherent stability of marine plants, originating from the rpm-dependence of the load. Note that $q_{E,1}$ is relatively small, as it is dependent upon engine friction losses, which are usually small compared to engine mechanical power delivery.
- $q_{E,2} \ge 0$: typically, when the engine is operating in a significantly large vicinity of steady-state operating points, it holds that $q_{E,2} \approx 0$, implying that the turbocharging system provides adequate combustion air flow rate and turbocharger speed variation does not have any impact on engine acceleration or deceleration. However, it can happen that turbocharger speed deviation (specifically, in the negative sense) is

introducing reduction in engine torque delivery as air-to-fuel ratio (A/F) is reduced to levels so that combustion is incomplete.

- $q_{TC,1} \geq 0$: the value of this derivative coefficient is usually small for two-stroke engines, such as the large Diesel engines used for ship propulsion. This parameter depicts the small, yet positive, effect on turbocharger accelerating torque of any small increase in engine speed. This is due to the effect on fuel mass flow rate (in kg/s, not the fuel index), which results in increased exhaust flow rate through the turbine and in effect increased turbine torque. Alternatively, the positive effect mentioned before can be attributed to faster compression, which results in increased values for maximum in-cylinder pressure and exhaust temperature, resulting in slightly increased turbine torque.
- $q_{TC\min,2} < 0$ and $q_{TC\max,2} \geq 0$: note that in a significantly large vicinity of steady-state operating points, $q_{TC,2} < 0$ and actually is quite close to $q_{TC\min,2}$. This is due to inherent turbocharger stability. As already demonstrated in Chapter 2, turbine torque is a roughly proportional function of turbo rpm, provided that fuel index and engine rpm are held constant. On the other hand, compressor load torque can be approximated as a polynomial (binomial, to be exact) function of turbo rpm. The two curves intersect at the value of turbo rpm that corresponds to the steady-state operating point, thus ensuring open-loop turbocharger stability.
- $q_{E\max,0} > 0$ and $q_{E\min,0} \leq 0$: note that in a significantly large vicinity of steady-state operating points, $q_{E,0} > 0$ and actually is quite close to $q_{E\max,0}$. This derivative coefficient is actually expressing the impact of fuel index on engine torque delivery. Therefore, it is positive due to the fact that, near steady-state, the engine can accelerate if the index is increased or decelerate if the index is reduced. However, under certain conditions $q_{E,0}$ may change sign. This can happen if the turbocharging system cannot sustain an adequate air flow rate for complete combustion. Therefore, for such low air flow rate values, any index increase results in an air-to-fuel (A/F) ratio reduction, leading to further engine torque delivery drop.
- $q_{TC,0} > 0$: this derivative coefficient is positive due to the fact that a fuel index increase results in turbocharger acceleration, and *vice versa* an index decrease results in turbo deceleration. This can be explained by taking into account that turbine torque depends upon exhaust flow rate, which is increasing when fuel flow rate is increased.

The above remarks for the partial derivatives coefficients justify the assumptions used in the calculation of the bounds of the elements of matrices A and B, in the linearised perturbation model of the typical marine plant. If one or more of the above assumptions, however, is not valid, then the bounds can still be determined, taking into account that extremum values are not lying in the interior of the value regions of the plant physical parameters.

Furthermore, for the application of robust control synthesis methods, bounding the disturbance vector $\mathrm{d} \in \mathbb{R}^2$ is needed. As already seen, only $d_1 \neq 0$. Specifically:

$$d_1 = -\frac{N_0^2 \cdot k_Q}{I} - \frac{\dot{N}_0 \cdot \Delta I}{I} = d_{\mathrm{ss}}(k_Q, N_0, I) + d_{\mathrm{tr}}(\dot{N}_0, I) \tag{5.91}$$

where obviously:

$$d_{\mathrm{ss}}(k_Q, N_0, I) = -\frac{N_0^2 \cdot k_Q}{I} \text{ and } d_{\mathrm{tr}}(\dot{N}_0, I) = -\dot{N}_0 \cdot \frac{\Delta I}{I} = \dot{N}_0 \cdot (\frac{I_0}{I} - 1) \tag{5.92}$$

The most convenient way to determine the extremum values $d_{\min}$ and $d_{\max}$ of the disturbance signal $d_1(t)$ is to determine the extremum values of each one of its two components, *i.e.* $d_{\mathrm{ss,min}}$ and $d_{\mathrm{ss,max}}$, as well as, $d_{\mathrm{tr,min}}$ and $d_{\mathrm{tr,max}}$.

First, $d_{\mathrm{ss}}(k_Q, N_0, I)$ can assume both negative and positive values. The central value is zero, corresponding to no deviation of the propeller torque coefficient K_Q from its nominal value K_{Q0}. In effect $d_{\mathrm{ss,min}} < 0$ and $d_{\mathrm{ss,max}} > 0$ corresponds to the minimum and maximum values of propeller torque coefficient perturbation signal k_Q, $(-\delta K_Q)$ and $(+\delta K_Q)$ respectively, where $\delta K_Q = \max|\Delta K_Q|$, a value dictated by screw propeller hydrodynamics. Assuming, without any loss of generality, $\max\{N_0\} = N_{MCR}$ where N_{MCR} is the MCR shaft speed, we obtain that:

$$d_{\mathrm{ss,min}} = -\frac{N_{MCR}^2 \cdot \delta K_Q}{I_0 - \delta I} \text{ and } d_{\mathrm{ss,max}} = +\frac{N_{MCR}^2 \cdot \delta K_Q}{I_0 - \delta I} \tag{5.93}$$

For the second component of the stochastic disturbance signal $d_1(t)$, *i.e.* $d_{\mathrm{tr}}(\dot{N}_0, I)$, it should first be noted that $d_{\mathrm{tr}} \equiv 0$ when $\dot{N}_0 = 0$, *i.e.* when no change in operating point occurs. However, as a nonzero value of $\dot{N}_0$ may correspond to an engine acceleration or deceleration, it can be concluded that:

$$\dot{N}_{0,\min} < 0 \text{ and } \dot{N}_{0,\max} > 0 \tag{5.94}$$

where $\dot{N}_{0,\min}$ and $\dot{N}_{0,\max}$ are obviously the extremum values of $\dot{N}_0$. These values can be determined from the equations of the N^{2}M. Indeed, as the number of operating point transitions is practically limited for marine plants, this is possible. In effect:

$$d_{\mathrm{tr,min}} = -\frac{\max\{(+\dot{N}_{0,\max} \cdot \delta I), (-\dot{N}_{0,\min} \cdot \delta I)\}}{I_0 - \delta I} \leq 0 \tag{5.95}$$

and

$$d_{\mathrm{tr,max}} = -\frac{\min\{(-\dot{N}_{0,\max}\cdot\delta I),(+\dot{N}_{0,\min}\cdot\delta I)\}}{I_0-\delta I} \geq 0 \tag{5.96}$$

However, it should be noted that operating point changes are performed rather slowly in common marine practice, *i.e.* $\dot{N}_0$ is assuming values close to zero. This, combined with the fact that the ratio $\delta I / I$ assumes values below 10–20%, gives values for this component of disturbance that:

$$|d_{\mathrm{tr}}| \ll |d_{\mathrm{ss}}| \tag{5.97}$$

Therefore, for all practical purposes, component d_{tr} can be either completely neglected or incorporated in d_{ss} by assuming a slightly increased value for the value of the uncertainty radius δK_Q of propeller torque coefficient.

5.4.3 Uncertainty Identification of Typical Powerplant

Using the neural torque approximator functions obtained previously for the test case of the MAN B&W 6L60MC propulsion plant, the ranges for the physical parameters are given in Table 5.4, as calculated by the procedure described above.

Table 5.4 Values of test case propulsion plant parametric uncertainties

$q_{E,1}$ (N m/rpm)	0.44–0.47	$q_{TC,1}$ (N m/rpm)	4.6–4.9
$q_{E,2}$ (N m/rpm)	0–185	$q_{TC,2}$ (N m/rpm)	-45–196
$q_{E,0}$ (kN m/%)	8.0–9.2	$q_{TC,0}$ (N m/%)	695–730
N_0 (rpm)	60–120		
K_Q (N m/rpm^2)	45.0–60.0		
I (10^3 kg m^2)	5.82–6.24		

The ranges determined above comply with the general remarks given for marine plants previously. The only exception is for parameter $q_{E,0} = \partial Q_E / \partial F_R$; indeed, note that the range provided for this is not as wide as mentioned in theory. This is due to the employment of the so-called "scavenging pressure-limiter" applied to the index value. This limiter function prevents the lack of combustion air mass (resulting in low A/F values), which causes the engine torque delivery to drop when the index is increased. Therefore, perfect combustion conditions are maintained by this rather obscure form of feedback control, thus reducing the range of parametric uncertainty for $q_{E,0} = \partial Q_E / \partial F_R$.

5.5 Transfer Function Matrix of the Marine Plant

5.5.1 The Open-loop Transfer Function Matrix

Similar to the procedure for the empirical transfer function, the transfer function matrix for the linear perturbation model will be derived in this section. The state equations of UPM in matrix form are:

$$\delta\dot{\mathrm{x}}(t) = \mathrm{A}\cdot\delta\mathrm{x}(t) + \mathrm{B}\cdot\delta u(t) + \mathrm{d}(t)$$

For convenience δ is dropped, keeping in mind that, in the case of the UPM, the state vector is actually the one holding the deviations of shaft and turbo rpm, *i.e.*:

$$\mathrm{x} = \begin{bmatrix} n & n_{TC} \end{bmatrix}^T = \begin{bmatrix} N - N_0 & N_{TC} - N_{TC0} \end{bmatrix}^\mathrm{T} \text{ and } u = f_R = F_R - F_{R0} \tag{5.98}$$

In effect:

$$\dot{\mathrm{x}}(t) = \mathrm{A}\cdot\mathrm{x}(t) + \mathrm{B}\cdot u(t) + \mathrm{d}(t) \tag{5.99}$$

By using the Laplace transform:

$$(s\mathrm{I}-\mathrm{A})\cdot\mathrm{x}(s) = \mathrm{B}\cdot u(s) + \mathrm{d}(s) \Leftrightarrow \mathrm{x}(s) = (s\mathrm{I}-\mathrm{A})^{-1}\cdot\mathrm{B}\cdot u(s) + (s\mathrm{I}-\mathrm{A})^{-1}\cdot\mathrm{d}(s) \tag{5.100}$$

In specific:

$$\mathrm{x}(s) = \mathrm{T_{xu}}(s)\cdot u(s) + \mathrm{T_{xd}}(s)\cdot\mathrm{d}(s) \tag{5.101}$$

where:

$$\mathrm{T_{xu}}(s) = (s\mathrm{I}-\mathrm{A})^{-1}\cdot\mathrm{B} = \frac{1}{p_0(s)}\cdot\begin{bmatrix} b_1 s - b_1 a_{22} + b_2 a_{12} \\ b_2 s + b_1 a_{21} - b_2 a_{11} \end{bmatrix} \tag{5.102}$$

$$\mathrm{T_{xd}}(s) = (s\mathrm{I}-\mathrm{A})^{-1} = \frac{1}{p_0(s)}\cdot\begin{bmatrix} s - a_{22} & a_{12} \\ a_{21} & s - a_{11} \end{bmatrix} \tag{5.103}$$

Finally, $p_0(s)$ is the open-loop characteristic polynomial given by:

$$p_0(s) = s^2 - (a_{11} + a_{22})\cdot s + (a_{11}a_{22} - a_{12}a_{21}) \tag{5.104}$$

It is evident that the parametric uncertainty, incorporated in the UPM state equations, presented previously, is appearing in the above transfer function matrices $\mathrm{T_{xu}}(s)$ and $\mathrm{T_{xd}}(s)$. This is examined in more detail in upcoming chapters. Here, another important aspect of typical marine Diesel powerplants is given attention.

The argumentation is based upon the following assumption valid for typical marine plants:

$$q_{E,2} = \frac{\partial Q_E}{\partial N_{TC}} \approx 0 \Rightarrow a_{12} = \frac{q_{E,2}}{I} \approx 0 \tag{5.105}$$

This assumption is justified by the fact that the turbocharging system is appropriately matched to the engine so that adequate combustion air is driven to the engine cylinders. Therefore, it can be assumed that engine torque is practically insensitive to turbocharger rpm variations for a significantly large vicinity of steady-state operating points. Taking into consideration this assumption, the open-loop characteristic polynomial becomes:

$$p_0(s) = s^2 - (a_{11} + a_{22}) \cdot s + a_{11}a_{22} = (s - a_{11}) \cdot (s - a_{22}) \tag{5.106}$$

Therefore, the open-loop poles of the propulsion system are $a_{11} = (q_{E,1} - 2K_{Q0}N_0)/I$ (slow pole) and $a_{22} = q_{TC,2}/I_{TC}$ (fast pole). Both poles are stable, at least over a wide range of values, according to the remarks presented in Section 5.4.3.

Combining the above assumption for a_{12} with the fact that only component d_1 of the disturbance vector is nonzero, the scalar open-loop transfer functions of interest are as follows:

$$T_{\mathrm{xu},1}(s) = \frac{n(s)}{f_R(s)} = \frac{b_1}{s - a_{11}}$$

and (5.107)

$$T_{\mathrm{xu},2}(s) = \frac{n_{TC}(s)}{f_R(s)} = \frac{b_2 s + b_1 a_{21} - b_2 a_{11}}{s^2 - (a_{11} + a_{22}) \cdot s + a_{11}a_{22}}$$

$$T_{\mathrm{xd},1}(s) = \frac{n(s)}{d_1(s)} = \frac{1}{s - a_{11}}$$

and (5.108)

$$T_{\mathrm{xd},2}(s) = \frac{n_{TC}(s)}{d_1(s)} = \frac{a_{21}}{s^2 - (a_{11} + a_{22}) \cdot s + a_{11}a_{22}}$$

Note that assumption $q_{E,2} = 0$ leads to zero-pole cancellation for the two scalar transfer functions from control action u and disturbance d_1 to shaft rpm fluctuation n. The physical interpretation for this is that, around a steady-state operating point, loading or unloading of the engine can be performed without any interference from the turbocharging system. This is why the pole cancelled is the fast one $(a_{22} = q_{TC,2}/I_{TC})$, which is introduced by the turboshaft dynamical equation.

5.5.2 Empirical and State-space Transfer Function

The correspondence between the transfer function obtained for $n(s)$ using empirical arguments (Chapter 3) and state-space analysis (Chapter 5) is investigated as a conclusion to this chapter. In Chapter 3 the following expression was obtained, referred to as a "reduced-order transfer function of the marine plant":

$$n(s) = \frac{C}{Is + 2K_{Q0}N_0} \cdot f_R(s) - \frac{N_0^2}{Is + 2K_{Q0}N_0} \cdot k_Q(s)$$

On the other hand, by using state-space analysis and linearisation for the marine plant the following expression has been obtained:

$$n(s) = T_{\text{xu},1}(s) \cdot f_R(s) + T_{\text{xd},1}(s) \cdot d_1(s) = \frac{b_1}{s - a_{11}} \cdot f_R(s) + \frac{1}{s - a_{11}} \cdot d_1(s)$$

By substituting in the above the values of a_{11}, b_1 and $d_1(s)$ (the latter as a function of $k_Q(s)$ only; the effect of rpm rate $\dot{N}_0$ and inertia uncertainty ΔI is not included, as the transfer functions of interest are the ones for the speed regulation problem) we obtain that:

$$n(s) = \frac{q_{E,0} / I}{s - \dfrac{q_{E,1} - 2K_{Q0}N_0}{I}} \cdot f_R(s) + \frac{1}{s - \dfrac{q_{E,1} - 2K_{Q0}N_0}{I}} \cdot \left(-\frac{N_0^2 \cdot k_Q(s)}{I} \right)$$

$$= \frac{q_{E,0}}{Is + (2K_{Q0}N_0 - q_{E,1})} \cdot f_R(s) - \frac{N_0^2}{Is + (2K_{Q0}N_0 - q_{E,1})} \cdot k_Q(s) \tag{5.109}$$

As can be seen, the coincidence of the two open-loop descriptions of the marine plant, *i.e.* the state-space approach and the empirical transfer function, is complete. Indeed, as said previously, $q_{E,1} = \partial Q_E / \partial N$ depends mainly upon the engine friction load and is comparatively small compared to the power absorption of the propeller, as expressed by the term $(2K_{Q0}N_0)$ in the transfer function. Therefore:

$$q_{E,1} \ll 2K_{Q0}N_0 \Rightarrow 2K_{Q0}N_0 - q_{E,1} \approx 2K_{Q0}N_0 \tag{5.110}$$

Finally, the significant notional coincidence between $q_{E,0}$ and parameter C should be noted. Indeed:

$$q_{E,0} = \left. \frac{\partial Q_E}{\partial F_R} \right|_{(N_0, N_{TC0}, F_{R0})}$$

Constant C is calculated in analogy from detailed thermodynamic models, such as MoTher, for a step fuel index change without any change in load torque and by artificially maintaining shaft rpm fixed (this is achieved by virtually introducing heavily increased shaft inertia). Turbocharger rpm can also be maintained fixed in the fuel index step experiment by increasing the turboshaft inertia; however, this is not considered necessary, except maybe when very large steps are applied, due to the independence of shaft acceleration from turboshaft rpm ($q_{E,2} \approx 0$). Therefore, the proposed procedure for determining thermodynamic gain C actually implements the calculation of the partial derivative $q_{E,0} = \partial Q_E / \partial F_R$ with the use of detailed control-volume thermodynamic models.

However, estimation of the exact value of the powerplant pole $a_{22} = q_{TC,2} / I_{TC}$ can also be performed using the fuel index step response of the engine in terms of torque delivery. Indeed, as demonstrated in Section 3.3.5, a first-order transfer function is fit to the transient torque response, the time constant of which τ_{TC}

corresponds to the fast pole of the system. This is possible from the system point of view, because constant $q_{E,2} = \partial Q_E / \partial N_{TC}$, however small, is not exactly zero; therefore, the zero-pole cancellation is not "complete", allowing some effect of the turbocharger dynamics to be introduced in the shaft rpm dynamical equation.

In any case, state-space dynamical analysis has proven a rather valuable theoretical tool, as it revealed the zero-pole cancellation, due to the small value of $q_{E,2} = \partial Q_E / \partial N_{TC}$. Therefore, the difficulty of identifying turbocharging time constant τ_{TC} should be attributed not only to its small value relative to τ_{prop}, as mentioned in Chapter 3, but also to the effect of zero-pole cancellation.

Furthermore, state-space analysis of the marine plant has fully justified the choice in using the reduced-order transfer function for controller design. Indeed, due to the zero-pole cancellation, the order of the characteristic polynomial of both transfer functions $T_{\mathrm{xu},1}(s)$ and $T_{\mathrm{xd},1}(s)$ involved in the dynamical equations for shaft rpm fluctuation n is formally reduced to one.

In conclusion, it can be said that state-space analysis of the marine plant enables for a deeper insight into its operation, as:

- The formulation of the non-linear state-space equations was performed on the basis of a widely accepted physical model, without the need for assumptions, thus improving confidence in the conclusions.
- No experiments are directly needed because the cycle-mean, quasi-steady engine model can be calibrated directly from manufacturer data.
- The notion of nominal (or normal) operation of the marine propulsion plant is clarified and "quantified" in the dynamical equations of the N^2M.
- The operating point regulation or tracking problem for ship propulsion powerplants is clearly stated on the basis of the (linear) UPM.
- The parametric uncertainties in the system matrices, appearing in the UPM state equations, can be safely determined over the entire operating range of interest using the neural torque approximators.
- The propeller load disturbance components are clearly determined in both qualitative and quantitative terms.

Therefore, any attempt to develop or improve control strategies for the marine engine and the propulsion system can use the state-space analysis method presented in this chapter, during design, as well as validation of the proposed controllers. This procedure is exemplified in the upcoming chapter for a full-state-feedback control scheme.

5.6 Summary

The state-space equations of the marine plant are formulated using the non-linear mapping abilities of neural nets. Instead of employing some standard functions for curve-fitting the torque maps, the automated calibration facilities of neural nets are preferred. This offers a series of advantages, the most important of which are independence from the type of engine and the ability to automatically manipulate incomplete and noisy training sets. However, a major disadvantage is that the complexity of the approximator torque functions obtained prohibits the use of analytical non-linear controller synthesis methods. Therefore, the state equations have to be linearised and decomposed to a nominal non-linear part and a locally linearised one. Then, in order to deal with the problem of different local linear problems, the linearised differential equations are assumed to have uncertain parameters, including the entire value sets. In effect, the uncertainty ranges are determined using the expressions of the torque approximator functions.

CHAPTER 6

MARINE PLANT ROBUST STATE-FEEDBACK CONTROL

6.1 Introduction

6.1.1. Controller Design Framework

It is well known that modern control techniques are based on the concept of state and state feedback in order to solve control problems such as disturbance rejection and setpoint tracking. The deployment of any state-feedback control scheme assumes that an open-loop plant description of the following form is available:

$$\dot{\mathrm{x}} = \mathrm{f}(\mathrm{x},\mathrm{u}) \tag{6.1}$$

where $\mathrm{x}^{\mathrm{T}}=[x_1\ x_2\ \dots\ x_n]$ is the n-dimensional state column vector, including all the dynamical variables of interest and $\mathrm{u}^{\mathrm{T}}=[u_1\ u_2\ \dots\ u_m]$ is the m-dimensional column vector of control actions applied to the system.

A more general form of the state equations will be considered here, including plant uncertainty and disturbance:

$$\dot{\mathrm{x}} = \mathrm{f}(\mathrm{x},\mathrm{u},\mathrm{\rho}) \tag{6.2}$$

Vector $\mathrm{\rho}^{\mathrm{T}}=[\rho_1\ \rho_2\ \dots\ \rho_l]$ is the l-dimensional column vector including all uncertain parameters of function f, *i.e.* sources of disturbance and uncertainty.

In the case of marine powerplants, such a state-space model has been obtained on the basis of physical principles and by employing neural nets in the previous chapter. As explained, the state model of the marine plant is of the form:

$$\begin{bmatrix} \dot{N} \\ \dot{N}_{TC} \end{bmatrix} = \begin{bmatrix} I^{-1}\cdot\left(Q_E(N,N_{TC},F_R)-K_Q\cdot N^2\right) \\ I_{TC}^{-1}\cdot Q_{TC}(N,N_{TC},F_R) \end{bmatrix} \tag{6.3}$$

where the engine torque delivery and turbocharger total torque are calculated using the neural approximator functions:

$$Q_E(N,N_{TC},F_R) = \mathrm{Q}_{\mathrm{Emax}}\cdot\{w_{00}+\sum_{i=1}^{7} w_{0i}\cdot\Phi(w_{b,i}+w_{E,i}\cdot N+w_{TC,i}\cdot N_{TC}+w_{R,i}\cdot F_R)\}$$

$$Q_{TC}(N,N_{TC},F_R) =$$

$$\mathrm{Q}_{\mathrm{Tmax}}\cdot\{v_{00}+\sum_{i=1}^{7} v_{0i}\cdot\Phi(v_{b,i}+v_{E,i}\cdot N+v_{TC,i}\cdot N_{TC}+v_{R,i}\cdot F_R)\} + \mathrm{Q}_{\mathrm{Cmax}}$$

Assuming a state-space description of the open-loop system is available, full-state feedback controls can be applied according to the following:

$$u = f_C(x) \tag{6.4}$$

in order to obtain a closed-loop system with the desired transient performance and steady-state error value.

In the case of marine plants, propeller load disturbance rejection and robustness against the uncertainty of shafting inertia are the most critical requirements for propulsion control system design. However, the complexity of the state transition function f, which has been formulated using the neural torque approximators, makes the direct synthesis of a controller of the above form rather difficult. Therefore, linearisation of the non-linear state equations is employed in order to deduce descriptions of the powerplant operation that are convenient for controller design.

Specifically, as analysed in the previous chapter, the propulsion system state-space non-linear description, repeated above for the sake of compactness, has been decomposed to the following two parts:

- the Nominal Non-linear Model (N^2M)
- the linear Uncertain Perturbation Model (UPM)

N^2M state equations are certain, but non-linear, differential equations. On the other hand, UPM results in the typical form of an uncertain linear system with exogenous disturbance. The control objectives for both the above models are investigated in the following sections.

6.1.2. Control of N^2M

The N^2M state equations are:

$$\begin{bmatrix} \dot{N}_0 \\ \dot{N}_{TC0} \end{bmatrix} = \begin{bmatrix} I_0^{-1} \cdot \left(Q_E(N_0, N_{TC0}, F_{R0}) - K_{Q0} \cdot N_0^2 \right) \\ I_{TC}^{-1} \cdot Q_{TC}(N_0, N_{TC0}, F_{R0}) \end{bmatrix} \tag{6.5}$$

where $Q_E(N_0, N_{TC0}, F_{R0})$ is the map of engine torque delivery and $Q_{TC}(N_0, N_{TC0}, F_{R0})$ is the map of turbocharger total (accelerating or decelerating) torque. The N^2M forms the reference model, representing the desired ideal manner of operation of the marine powerplant.

Two discrete modes of operation can be distinguished for N^2M:

- <u>Steady-state operation</u>. In this mode $\dot{N}_0 = \dot{N}_{TC0} = 0$, which is translated to the following equations for engine and turbo rpm, N_0 and N_{TC0} respectively, as well as, for the fuel index value, F_{R0}:

$$Q_E(N_0, N_{TC0}, F_{R0}) = K_{Q0} \cdot N_0^2 \tag{6.6}$$

$$Q_{TC}(N_0, N_{TC0}, F_{R0}) = 0 \tag{6.7}$$

- <u>Change of operating point</u>. In this mode it generally holds that $\dot{N}_0 \neq 0$ and $\dot{N}_{TC0} \neq 0$, meaning that a plant acceleration or deceleration is taking place or, equivalently, change of power level as percentage fraction of

MCR power is applied. In common marine practice such changes of power level, though, are performed using ramps of rather small slope, meaning that the values for $\dot{N}_0$ and $\dot{N}_{TC0}$ are maintained small; the only exceptions are emergency situations, such as crash stop and emergency reversing engine manoeuvres.

The engine manufacturer commonly provides all data and information regarding steady-state operation of the engine and the plant under nominal conditions, *i.e.* under the nominal propeller curve and for fixed shafting system inertia. Therefore, values of the triad (N_0, N_{TC0}, F_{R0}) for which the above equilibrium conditions hold can be readily determined from the manufacturer sheets. Specifically, engine brake power as a fraction of MCR power, $P_{\mathrm{SET\%}} = 100\% \cdot \left(Q_{L0} \cdot 2\pi \cdot N_0 / 60\right) / P_{MCR}$, and rpm, N_0, at each steady-state point are provided on a power–rpm chart, allowing thus the determination of the engine brake torque $Q_{E0} = Q_{L0} = K_{Q0} \cdot N_0^2$. Also note that, as explained in Section 3.3.4, a linear relationship of the form $F_{R0} = \kappa'_{\mathrm{brake},1} \cdot P_{\mathrm{SET\%}} + \kappa'_{\mathrm{brake},0}$ links the rated fuel index value F_{R0} with the rated engine power level setpoint $P_{\mathrm{SET\%}}$. Furthermore, the nominal propeller curve is projected on the compressor map, allowing the determination of scavenging pressure, turbocharger speed (N_{TC0}) and air flow rate at each steady-state operating point. In conclusion, each steady-state operating point of a marine plant is uniquely defined by the power level setpoint as a fraction of MCR, $P_{\mathrm{SET\%}}$.

A change in operating point is performed by applying a time schedule $P_{\mathrm{SET\%}}(t)$, most commonly of ramp or step form, to the plant. This power level schedule is translated to a fuel index schedule $F_{R0}(t)$ through the linear relationship given above. Assuming that the starting point is a steady-state operating point, too, the change of value for F_{R0} perturbs the balance in both the engine–propeller shaft and turbocharger shaft. For example, if F_{R0} is slightly increased from its current steady-state value, both $\dot{N}_0$ and $\dot{N}_{TC0}$ are expected to assume temporarily small positive values, resulting in slightly increased values for N_0 and N_{TC0}. Then, if no further increase in F_{R0} is applied, the plant is expected to reach a new point of equilibrium where the two equilibrium (for engine–propeller and turbine–compressor torques) conditions hold. The new operating point must also be a (higher) steady-state point on the propeller curve, otherwise the equilibrium conditions cannot be satisfied.

In both cases the N^2M, *i.e.* steady-state or change of operating point, can serve as a setpoint generator for the marine plant state variables, shaft and turbo rpm, N and N_{TC} respectively. In control terminology, the N^2M will form the core of a supervisory controller. It takes as input the rated power level, $P_{SET\%}$, and calculates the fuel index value F_{R0} for which the engine can deliver the required power under nominal conditions, as well as the corresponding setpoint (reference values) of the state variables, N_0 and N_{TC0}. Aspects of such a marine plant supervisory controller are examined closer later in this chapter.

6.1.3. Control of UPM

As demonstrated in the previous chapter, linearisation of the non-linear state equations of the marine plant provides the following linear state equations of a SIMO system with disturbance for the UPM:

$$\dot{\mathrm{x}} = \mathrm{A}\cdot \mathrm{x} + \mathrm{B}\cdot u + \mathrm{d}$$

where the state vector is $\mathrm{x}^{\mathrm{T}} = [n\ n_{TC}] = [(N - N_0)\ (N_{TC} - N_{TC0})]$, the control action is $u \equiv f_R = F_R - F_{R0}$ and the disturbance vector is $\mathrm{d}^{\mathrm{T}} = \left[-I^{-1}\cdot\left(N_0^2 \cdot k_Q + \dot{N}_0 \cdot \Delta I\right)\ \ 0\right]$. Furthermore, UPM matrices A and B are:

$$\mathrm{A} = \begin{bmatrix} \dfrac{q_{E,1} - 2K_{Q0}N_0}{I} & \dfrac{q_{E,2}}{I} \\ \dfrac{q_{TC,1}}{I_{TC}} & \dfrac{q_{TC,2}}{I_{TC}} \end{bmatrix} \text{ and } \mathrm{B} = \begin{bmatrix} \dfrac{q_{E,0}}{I} \\ \dfrac{q_{TC,0}}{I_{TC}} \end{bmatrix}$$

As explained in Section 5.4.2, the above matrices include bounded uncertainty and it holds that:

$$\mathrm{A} = \mathrm{A}_0 + \Delta\mathrm{A} \text{ and } \mathrm{B} = \mathrm{B}_0 + \Delta\mathrm{B} \tag{6.8}$$

where the nominal perturbation matrices, $\mathrm{A}_0 = [a_{ij,0}]$ and $\mathrm{B}_0 = [b_{i,0}]$, as well as intervals for the elements of matrices $\Delta\mathrm{A} = [\Delta a_{ij}]$ and $\Delta\mathrm{B} = [\Delta b_i]$, are calculated using matrices $\mathrm{A}_{\text{min or max}} = [a_{ij,\text{min or max}}]$ and $\mathrm{B}_{\text{min or max}} = [b_{i,\text{min or max}}]$, given in Section 5.4.2, through the following relationships:

$$a_{ij,0} = \frac{a_{ij,\min} + a_{ij,\max}}{2},\quad b_{i,0} = \frac{b_{i,\min} + b_{i,\max}}{2} \tag{6.9}$$

$$\left|\Delta a_{ij}\right| \leq \frac{a_{ij,\max} - a_{ij,\min}}{2},\quad \left|\Delta b_i\right| \leq \frac{b_{i,\max} - b_{i,\min}}{2} \tag{6.10}$$

As explained in Chapter 5, UPM expresses the deviation, in terms of operating point, of the actual plant from its ideal behaviour, quantified by the $\mathrm{N^2M}$ equations. The cause of this deviation, recorded in the UPM state vector $[n\ n_{TC}]^{\mathrm{T}}$, is the fluctuation of the propeller torque coefficient, k_Q, and the shafting system inertia, ΔI.

The control problem for the UPM is, in effect, to bound the operating point of the actual plant in a small vicinity of the ideal behaviour, as set by the value of the

N^2M state vector $[N_0\ N_{TC0}]^T$. In other words, for the UPM state vector, the requirement is translated into the following two inequalities:

$$|n| \leq \delta n \text{ and } |n_{TC}| \leq \delta n_{TC} \tag{6.11}$$

provided that the fluctuation of propeller torque, k_Q, and shafting system inertia, ΔI, are also bounded. An additional requirement for the stochastic nature of these signals is to have zero mean value, as no integral control is employed. This is also analysed further below.

As the state equations of UPM are linear, the closed-loop performance of UPM, coupled with linear full state feedback controls with constant gains of the form:

$$u \equiv f_R = \mathrm{K} \cdot \mathrm{x} = [K_1 \quad K_2] \cdot [n \quad n_{TC}]^T = K_1 \cdot n + K_2 \cdot n_{TC} \tag{6.12}$$

is investigated in this text. The conclusions made are rather important, especially concerning the inclusion of the second state variable of the marine plant, *i.e.* turboshaft rpm, which does not appear in the PID control law examined in Chapter 4. Indeed, as demonstrated below, employment of the second state variable allows a more complete approach to the solution of the control problem, thus reducing the need of empirical add-on modules, such as scavenging pressure limiter. At the same time, smoother engine running, as well as increased safety and reliability, of the powerplant is achieved.

6.1.4. Architecture of the Propulsion Control System

Based on the considerations presented for both N^2M and UPM control options, a complete engine and propulsion control system may consist of the structure presented in Figure 6.1.

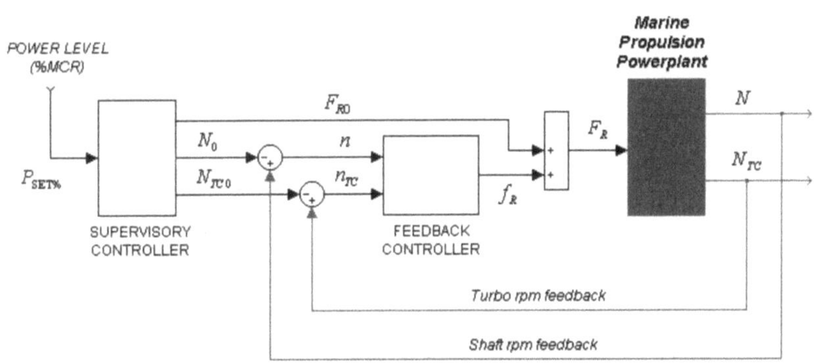

Figure 6.1 Propulsion control system architecture based on plant model decomposition

As already mentioned, the control system consists of a supervisory and a feedback controller following the partitioning of the non-linear state equations into a nominal part and a linearised part. According to the concept in Figure 6.1, the supervisory controller incorporates and "solves" in real-time the state equations of N^2M, in order to calculate the operating point setpoints to be fed to the feedback controller. On the other hand, the feedback controller maintains the operating point of the actual powerplant, which is subject to parametric uncertainty and disturbance, as close as required to the value of the N^2M. This is achieved by proper loop-shaping of the UPM, using a linear full-state feedback control law for rejecting propeller torque coefficient disturbance. Finally, note that robustness against parametric uncertainty introduced to the closed-loop system, due either to the linearisation procedure or the shafting inertia fluctuation, must also be considered during the feedback controller synthesis.

It should be mentioned here the PID speed governor, investigated in Chapter 4, or some other linear, or even non-linear, control scheme can be used as feedback controller of the marine plant. Each one of the possible feedback controllers must, however, be examined and evaluated on whether it can meet the design requirements or not. For example, any PI or PID control law gives zero steady-state error, whilst, as shown later, the state feedback controller can provide zero steady-state error only if the disturbance has zero temporal average. On the other hand, the PID control law accounts only for shaft rpm deviation and increases the closed-loop system's order by one. A comparative assessment of the PID control law with full-state feedback control is provided at the end of the present chapter, examining, as well, possible extensions for the combination of the two approaches for improved feedback control of the marine plant.

Finally, it is worth mentioning that the approach above is similar to that of Quantitative Feedback Theory (QFT) [51]. According to QFT, the control structure shown in Figure 5.2can be used.

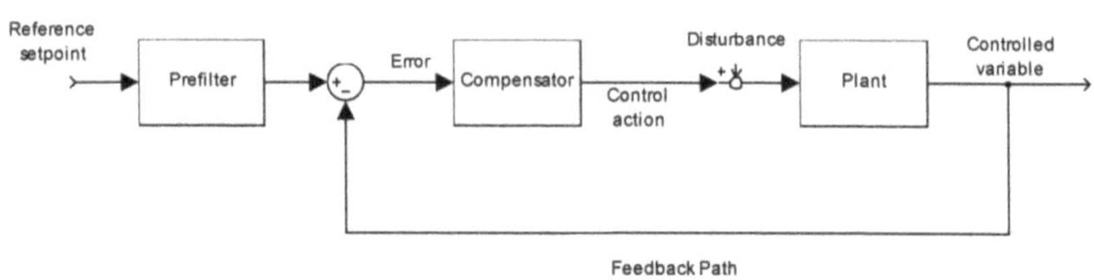

Figure 6.2 The QFT control scheme for linear systems

The analogy with the control structure proposed for the case of marine plants is completed, as the supervisory controller is the counterpart of the prefilter and the feedback controller is the counterpart of the compensator. In the same sense, the prefilter defines how fast a (step) setpoint change is applied to the closed-loop, whilst the compensator reduces the effect of the disturbance; robustness considerations for the resulting control system are also taken into account.

However, despite the generic similarity in control structure, there are significant differences, the most important of which are: (a) the generic QFT methodology refers to overall linear plants "in-the-large" and not plants described by locally linearised models derived from "global" non-linear ones; (b) QFT design requires an open-loop description of the plant with transfer function(s) and assumes, at least in the most trivial cases, that the plant is observable and controllable; (c) when QFT design methods are extended to multi-output systems, mathematical intricacy becomes quite cumbersome; therefore, for systems with just two outputs, such as the marine plant, there is a question of whether a method with comparable results but less algebraic implications can be found in order to avoid the formal synthesis procedure.

Based on the above, the analogy with the QFT approach will not be considered any more. Indeed, mainly due to (a) and (b), QFT is not directly applicable to marine plant control; as explained in Section 2.5.3, marine plant operation is highly non-linear when large operating point changes are taking place.

6.2 Supervisory Setpoint Control of the Marine Plant

6.2.1 Setpoint Control Requirements

Supervisory control of the marine propulsion powerplant, in the sense of setpoint generation for the feedback controller, forms a very important part of a complete marine engine and propulsion control system. In Chapter 4, for example, the control action f_R was assumed to be superimposed on a steady-state control F_{R0}, which was determined using the nominal propeller curve equation, in order to equate engine torque delivery with propeller torque demand, as follows:

$$C \cdot F_{R0} = K_{Q0} \cdot N_0^2$$

A number of difficulties exist, however, especially in the determination of thermodynamic gain C already mentioned in Chapter 3. This difficulty, combined with the one encountered for turbocharging time constant τ_{TC}, has motivated the use of thermodynamic models of engine operation. Using the quasi-steady, cycle-mean engine model of Chapter 2, the state-space model developed in Chapter 5 enabled us to substitute the above relation for steady-state engine torque delivery with the more accurate one:

$$Q_{E0} = Q_E(N_0, N_{TC0}, F_{R0}) = K_{Q0} \cdot N_0^2$$

In any case, the engine–propeller shaft equilibrium equation is straightforward at steady-state points provided by the manufacturer. When a deliberate engine acceleration or deceleration takes place, however, the aforementioned equilibrium and the equilibrium at the turboshaft are perturbed. In effect, as explained in Chapter 5, $\dot{N}_0$ and $\dot{N}_{TC0}$ assume nonzero (positive or negative) values. However, this will appear as error in the feedback controller, which will have to assume compensatory contol action. In effect, a major objective in introducing supervisory control is to make any deliberate change of operating point undetectable to the plant feedback controller, so that no excessive control action is generated and zero-steady state error is introduced. To achieve this it is imperative that well-specified dynamics is introduced in (shaft and turbo) rpm setpoint changes. Such dynamics can maintain deviations n and n_{TC} as close to zero as possible when deliberate changes in the operating point of the plant occur and no deviation from the nominal behaviour is observed.

Furthermore, as demonstrated in Chapter 4, P-control may give steady-state error, *i.e.* shaft rpm will deviate from the new required setpoint; later on in this chapter, this result is extended to the case where linear full-state-feedback control is used. This undesired effect can only be avoided, in either case of P-control or state-feedback control, by introducing an appropriate modification in the offset fuel index value F_{R0}, combined with changes in shaft rpm setpoint N_0, and turbo rpm setpoint N_{TC0}.

It is also argued that supervisory setpoint control may be needed in the case where an I-term is present in the control law. The I-term, as explained in detail in Section 4.1.3, has the virtue of eventually eliminating any steady-state error, due to the introduction of a closed-loop zero at $s = 0$; therefore, the need of supervisory control of the offset fuel index value F_{R0} is not direct as in the case of P-control or state feedback. However, it is still better to effectuate operating point changes through a supervisory controller because of the following reasons:

- If the PI(D) controller is tuned for disturbance rejection, then it usually obtains large, in the absolute sense, gains, in order to provide fast response. However, this means that the incremental control action f_R assumes large values for small values of n. Therefore, a setpoint prefilter (as proposed in the QFT control structure) is required that can discriminate the error values, and allow fast response when error is introduced due to disturbance. At the same time, it can moderate control intervention when the error is due to operating point change.
- Steady-state error elimination using the I-term is effectuated through the accumulation of error. Therefore, if it is used for operating point changes, large integral error values can be expected. This is translated to increased risk of integrator saturation. Furthermore, a possible integrator

reset introduces large (even prohibitive) discontinuity in the control action, making such an operation problematic, even if required.

Based on the considerations above, it is concluded that even if an I-term is included in the control law, it is preferable to effectuate operating point changes through a supervisory controller module.

However, it should be noted out that in common marine practice supervisory propulsion control is not used and changes in the operating point are effectuated through the PI speed governor. In effect, tuning of the governor for increased disturbance rejection is problematic. On the other hand, the supervisory controller, as proposed in this text, requires a real-time model of the powerplant so that the appropriate rpm setpoint temporal profiles are generated during operating point changes. This, as well as the positive effect of introducing a supervisory controller for the marine plant, is further investigated in Sections 6.2.2 and 6.2.3.

6.2.2 Supervisory Controller Structure

As already mentioned, the marine plant supervisory controller will be based on the N^2M equations. This approach applies the "nominal" plant dynamics to the engine and turbo rpm setpoints; due to the integrators, corresponding to the integrating action of the two shafts, first-order filtering is effectively applied to the two setpoints. However, the similarity to the QFT control structure ends here. Indeed, N^2M equations form a state-space realisation of a non-linear first-order filter, whilst the setpoint prefilter, used in QFT designs, has to be a linear one. Furthermore, note that as the feedback controller of the marine plant generates only the incremental control action f_R, the supervisory controller also has to generate the offset fuel index value F_{R0}. The proposed architecture of the supervisory controller module, based on N^2M, is shown in Figure 6.3.

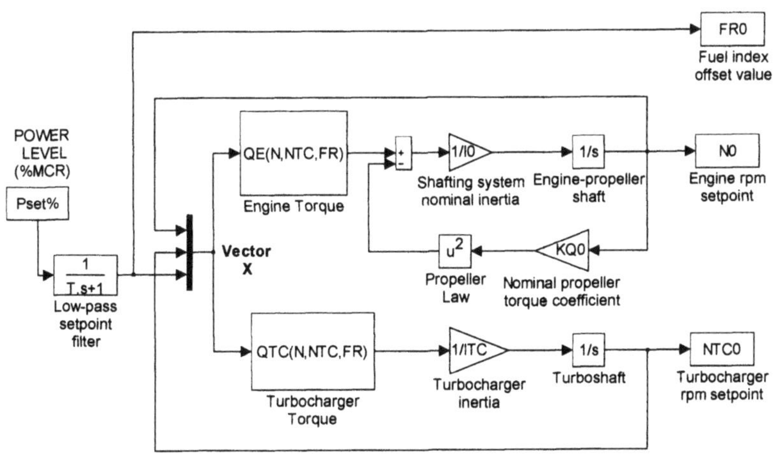

Figure 6.3 The N^2M in block diagram form for supervisory control applications

As can be seen in Figure 6.3, an additional dynamical element, namely a low-pass setpoint filter, has been added. This is a first-order linear filter with time constant τ_f, and as explained in the next paragraph, smoothens the effect of rapid large operating point changes on engine torque delivery. Until its purpose is clearly explained, however, it can be assumed that $\tau_f = 0$, meaning that the filter is temporarily, for the sake of simplicity, omitted.

Some important application aspects concerning supervisory controller implementation will now be examined. These aspects are mainly related to the digital (discrete-time) implementation of the N^2M equations in an actual shipboard supervisory controller module of marine plants. Namely:

- sampling and signal reconstruction
- integration method
- real-time torque calculation

Sampling frequency and period in marine plants are determined on the basis of engine crankshaft rotation. Because N^2M allows for cycle-mean estimation of variables, it is meaningful to choose the calculation sampling frequency equal to or larger than engine speed at MCR, *i.e.* $f_s \geq N_{MCR}/(60\text{ s/min})$, so that the calculation time step corresponds to one full engine cycle. As the practical operational speed range of directly coupled marine engines is between 30 and 120 rpm if sampling period is set to *e.g.* $T_s = (60\text{ s/min})/120\text{ rpm} = 0.5\text{ s}$, then in the low-speed area (*e.g.* 30–60 rpm) the temporal resolution effectively lies between two and four samples per cycle. Therefore, if no precautions are taken, it is possible that in-cycle variation of fuel index value may be induced, resulting in potential shafting resonance or other undesired effects. However, the frequency content of the fuel index signal is also determined by the continuous signal reconstruction technique used (Digital-to-Analogue, D/A conversion); for example, Zero-Order-Hold (ZOH) introduces stepwise changes in the continuous time signal. In conclusion, for certain applications, first-order hold might also be considered. Also, note that the first-order low-pass setpoint filter has a positive effect on this aspect of supervisory controller operation.

Related to the sampling frequency and the A/D–D/A conversion techniques used, is the integration method employed, *i.e.* the discrete-time implementation of the two integrators. In control practice, the most common approach is backwards difference, according to which the continuous-time integrator block shown in Figure 6.4 is substituted by the following discrete-time difference equation and corresponding transfer function (obtained using the Z-transform):

$$y(nT_s) - y\left((n-1)T_s\right) = T_s \cdot u(nT_s) \overset{Z}{\Leftrightarrow} \frac{Y(z)}{U(z)} = \frac{T_s}{1-z^{-1}} = \frac{T_s \cdot z}{z-1} \tag{6.13}$$

implemented by the (discrete-time) block diagram shown in Figure 6.5

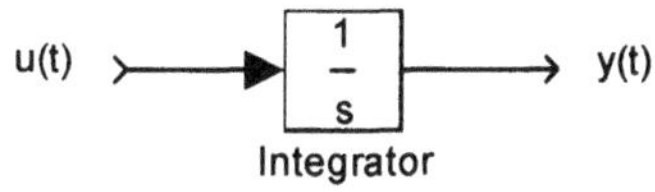

Figure 6.4 Continuous-time integrator in block form

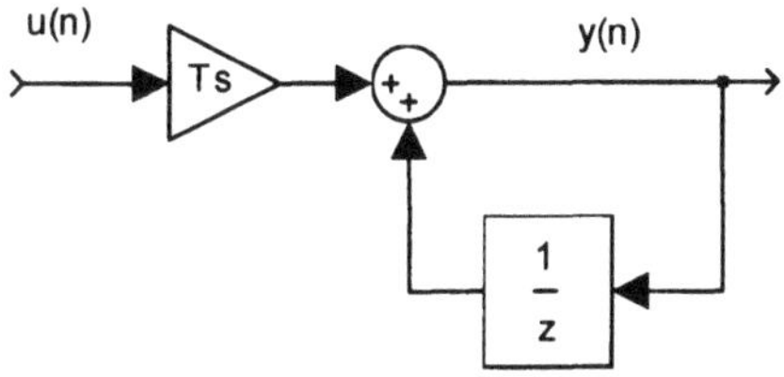

Figure 6.5 Discrete-time integration with backwards difference

Finally, the most important practical aspect of the marine supervisory controller is the form that will be used for the implementation of the engine and turbocharger torque functions, $Q_E(N, N_{TC}, F_R)$ and $Q_{TC}(N, N_{TC}, F_R)$. In Chapters 2 and 5 the following three alternative ways were presented for the determination of the torque variables:

- as solutions of a non-linear, perplexed algebraic system (see Sections 2.5.1 and 2.5.2)
- as 3D numerical maps with fixed step in Section 2.5.4
- as neural nets trained to approximate the numerical maps (see Section 5.2)

The first method cannot, of course, be considered for real-time implementation, as the computational overhead introduced by any numerical iterative solution of the algebraic system is regarded as prohibitive. On the other hand, the torque functions can be implemented as look-up tables or neural approximators. Although, the look-up tables implementation has an advantage because it does not require any kind of calibration, such as training for neural nets, there are a number of arguments supporting the choice of the neural approximators; these arguments are examined briefly below.

First, it has been argued [42,49] that neural nets are computationally more efficient implementations of a map than a look-up table. The computational resources in terms of either memory or number of operations required are usually reduced when neural nets substitute look-up tables, especially with a large number of dimensions (more than one). Moreover, commercial availability of parallel architectures based on the neural computing model, instead of a conventional von Neumann machine, makes this argument even stronger.

Training of the neural approximators allows handling of sparse, incomplete and noisy raw data sets in a better way; in contrast, if these data series are incorporated as look-up tables significant numerical error may be introduced. Note that in the case of marine plants, a number of variables, such as pressure and temperature of lubrication oil, and ambient (atmospheric) conditions, which affect steady-state performance, have been omitted from thermodynamic modelling. Their effect on performance, although small, is not completely negligible. However, a theoretical, physical model from first principles depicting this effect is difficult to construct. Furthermore, experimental results are not easily found and vary greatly from one plant to another or over long time periods for the same plant. Therefore, tuning of the nominal plant model, embedded in the supervisory controller, must be carried out while in service and whenever specific operating conditions are encountered. This operational aspect of any marine supervisory controller makes neural nets a rather attractive approach, because they can be re-trained whenever additional data are available, and form a convenient scheme of associative, cumulative memory [52–54].

A final aspect of the use of the neural torque approximators has to do with the robustness properties of the feedback controller of the marine plant. Specifically, as shown in Chapter 5, the parametric uncertainty of the UPM system matrices A and B depends upon inertial uncertainty ΔI, as well as uncertainty in the engine and turbocharger torque derivatives $q_{E,i}$ and $q_{TC,i}$, $i = 1,2$. The latter, however, are calculated using the neural torque approximators; to be specific, their analytical expressions can be calculated around each steady-state operating point, defined by a triad (N_0, N_{TC0}, F_{R0}), if the weight matrices ~~W~~, W_0 and ~~V~~, V_0 of the two approximators are known. Therefore, if the approximators, incorporated in the supervisory controller are retrained, uncertainty in UPM system matrices can be determined in an automated manner, too, using the new values of the elements of the weight matrices. Therefore, robustness assessment of the feedback controller can be effectuated automatically, whenever the nominal plant dynamics, as expressed by the parameters in the equations of N^2M, are detected to change.

The properties of the neural torque approximators for the implementation of the torque functions in a marine plant supervisory controller are investigated in the following test case of the MAN-B&W 6L60MC propulsion plant introduced in Chapter 5.

6.2.3 Test Case Investigation

The supervisory controller module for the MAN-B&W 6L60MC plant, whose open-loop thermodynamic behaviour was examined in Section 2.5.3, and the neural torque approximators for which were deduced in Section 5.2.3, is now investigated numerically.

Specifically, the dynamical response of the engine thermodynamic model of Chapter 2 is compared to the response of the system with the neural torque approximators. The transients examined are a 35 to 100% step and a 50 to 100% step in fuel index value. Note that no low-pass setpoint filter is placed between the power level setpoint and the fuel index offset value; moreover, for the sake of simplicity, the power level setpoint is assumed to coincide with the rated fuel index value, *i.e.* in the relation $F_{R0} = \kappa'_{\text{brake},1} \cdot P_{\text{SET\%}} + \kappa'_{\text{brake},0}$ it is assumed that $\kappa'_{\text{brake},1} = 1.0$ and $\kappa'_{\text{brake},0} = 0.0$. The shaft and turbo rpm responses, as well as engine torque delivery and turbocharger total torque, are plotted in Figures 6.6 and 6.7 for the two transient cases. Also, note that for the first case (35 to 100% step) the series generated by the thermodynamic model and the neural state model are both plotted. On the other hand, for the second case (50 to 100% step) only the difference (thermodynamic minus neural) is plotted for each variable; otherwise the plots would have been confusing as the modelling error assumes remarkably small values.

Discrete-time integration has been used with both models; the time step was 0.1 s. In that manner, simulation is closer to reality as explained previously. Also, as a side effect, simulation time in Matlab/Simulink is reduced significantly when a fixed integration time step is employed. Note that the value of the time step is adequately small. Indeed, as explained in Section 2.5.3, MAN-B&W suggest assigning an rpm value at MCR not larger than 123 rpm for the 6L60MC engine. Therefore, the sampling interval has to be less than or equal to:

$$T_s \leq (60\ \text{s/min})/123\ \text{rpm} = 0.488\ \text{s}$$

Therefore, the value of 0.1 s is approximately five times smaller and well below the above sampling threshold required for cycle-mean modelling temporal resolution.

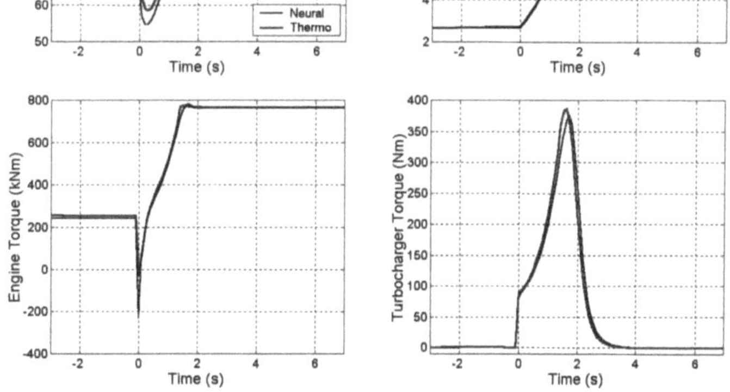

Figure 6.6 Transient 1: 35 to 100% load – Step response predicted by both models

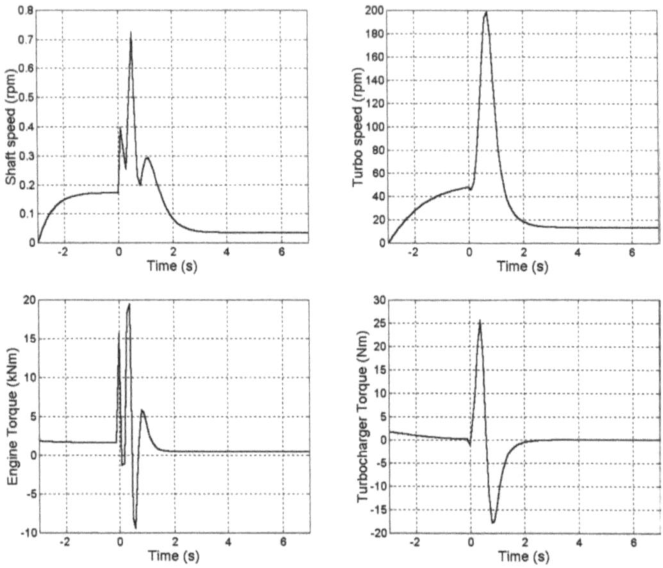

Figure 6.7 Transient 2: 50 to 100% load – modelling error (thermodynamic – neural)

It can be seen from the plots in Figures 6.6 and 6.7 that, if the response of the thermodynamic model is assumed to be the actual engine response, then the neural state model, which is based on the neural torque approximators, provides adequately small modelling error values. The modelling error is clearly increasing with the magnitude of the step; this is expected, as both models are non-linear and the magnitude of the excitation plays a rather important role. Note that a step change in load may be considered as a worst-case scenario of rapid engine loading. However, note that step engine load changes are very rarely applied in actual marine practice.

Finally, bearing in mind that some modelling errors exist in the thermodynamic engine model, it is reasonable also to expect that tuning (*i.e.* training) of the neural torque approximators will be needed in any actual installation. Also, after some time interval in service has elapsed, or after overhaul, it might be required that the torque approximators are re-trained so that their predictions are closer to the response of the actual plant.

6.2.4 The Low-pass Setpoint Filter

As already explained, a setpoint filter is placed in the supervisory controller structure. The objective in introducing this filter is to smoothen very rapid changes of engine loading. This is required for a variety of reasons, such as thermostructural strength of the plant components, shafting system resonance, turbocharging system lag and pulsing phenomena, *etc.* All these may cause severe engine and/or powerplant failures of various kinds. Therefore, limitation on the rate of engine loading change is required. One way to achieve this is by low-pass filtering the setpoint; evidently, other methods exist, such as slew-rate limiters, *etc*.

As the physical engine model used throughout this work covers in adequate detail the thermodynamic viewpoint of the powerplant operation, the only limitation that will be considered in this text is turbocharging system lag. Specifically, as can be seen from Figures 2.3–2.6, 6.6 and 6.7, when a large fuel index step is applied the engine torque delivery temporarily fails to follow up and, as a consequence, a temporary shaft deceleration is observed. This is due to the sudden increase of exhaust pressure that causes reduced air flow through the engine; as shown in the Figures 2.3–2.6 the air mass flow rate is temporarily decreased. In effect, A/F is reduced to values lower than $(A/F)_{low}$. Eventually, however, the increased energetic content of the exhaust gas causes the turbocharger to accelerate, therefore increasing the scavenging pressure, the air flow rate and the engine torque delivery.

Problems like the above can be dealt with using non-linear numerical techniques. As the specifics vary from one case to another the methodology is exemplified through the test case of the MAN-B&W 6L60MC engine, which is considered as a typical case of a ship propulsion powerplant. In this test case, the main problem is to determine the time constant τ_f of the first-order low-pass filter with transfer function:

$$H_f(s) = \frac{1}{1+\tau_f \cdot s} \tag{6.14}$$

The objective will be to filter the fuel index (or equivalently the power level setpoint) signal so that no deceleration of the engine–propeller shaft is observed during the transient 35 to 100%. As can be seen in Figures 2.3–2.6, 6.6 and 6.7, this is a worst-case scenario of engine loading, where significant shaft rpm reduction is observed in the first second of the transient. For the investigation, the neural state model will be used for design; however, validation will include the thermodynamic model, too.

In both cases the discrete-time integration with the backwards difference method is used. Therefore, the filter transfer function is transformed in the Z-domain, where its representation is:

$$H_f(z) = H_f(s)\Big|_{s=\frac{1-z^{-1}}{T_s}} = \frac{T_s}{(T_s+\tau_f)-\tau_f \cdot z^{-1}} = \frac{T_s \cdot z}{(T_s+\tau_f)\cdot z-\tau_f} \tag{6.15}$$

where $T_s = 0.1\,\mathrm{s}$ is the sampling period determined in Section 6.2.3. Taking into account Shannon's sampling theorem, we obtain that a time constant value $\tau_f < 2 \cdot T_s = 0.2\,\mathrm{s}$ is meaningless.

The effect of the filter can be understood by reference to Figure 6.8. The plant response, in terms of shaft rpm, for the 35 to 100% transient is plotted with two different values of τ_f : 0.25 and 1.0 s. As can be seen, although the presence of a filter with small time constant (just 2.5 times larger than Shannon's limit) improves the rpm response greatly, shaft deceleration can still be observed. On the other hand, when the time constant is large then engine acceleration is delayed without reason.

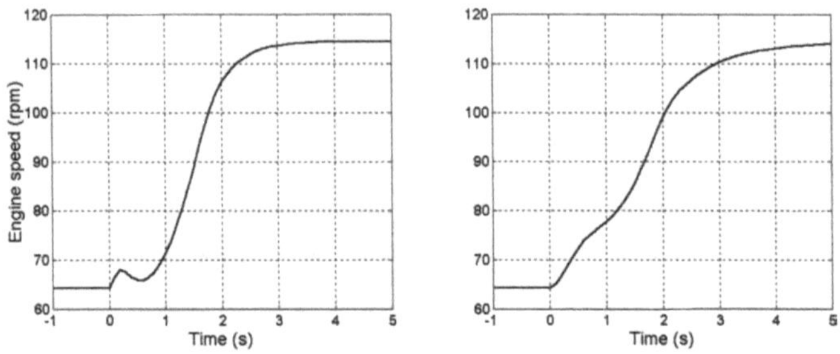

Figure 6.8 Effect of filter on response: $\tau_f = 0.25\,\text{s}$ (left) and $\tau_f = 1.0\,\text{s}$ (right)

In this respect, a trial-and-error search for the value of τ_f has been performed in the interval $[0.25\,\text{s}, 1.0\,\text{s}]$ with a $10^{-2}\,\text{s}$ step, in order to obtain the value that gives $\dot{N}_0 \geq 0$ (no shaft deceleration condition) during the entire transient 35 to 100%. The "optimum" time constant was $\tau_f = 0.47$ s, for which the plant response is given in Figure 6.9 using both the neural state model and the thermodynamic engine model.

As can be seen, the neural state model provides a response very close to that of the thermodynamic engine model, supporting the choice of using the neural approximators for the filter design. On the other hand, the neural state model required less execution time on the computer, compared to the thermodynamic model, enabling a rapid development and more complete coverage of the time constant value set.

The conclusion drawn from Figure 6.9 is that the plant is accelerating in a smoother manner when the setpoint filter is used. On the other hand, note that because the system is non-linear the duration of the transient is only slightly increased (no more than 0.5 s) compared to the no filter case. The smoother acceleration is also manifested by the profile of engine torque delivery, as predicted by the reference thermodynamic engine model. Note that although this signal still exhibits a drop, the engine torque value does not fall below the value at 35% load. On the other hand, when no filter is used, the engine torque exhibits a large reduction, well below the value at 35% load. Finally, for reference purposes, the filtered fuel index schedule, applied to both the thermodynamic engine model and the neural state model, is also plotted in Figure 6.9.

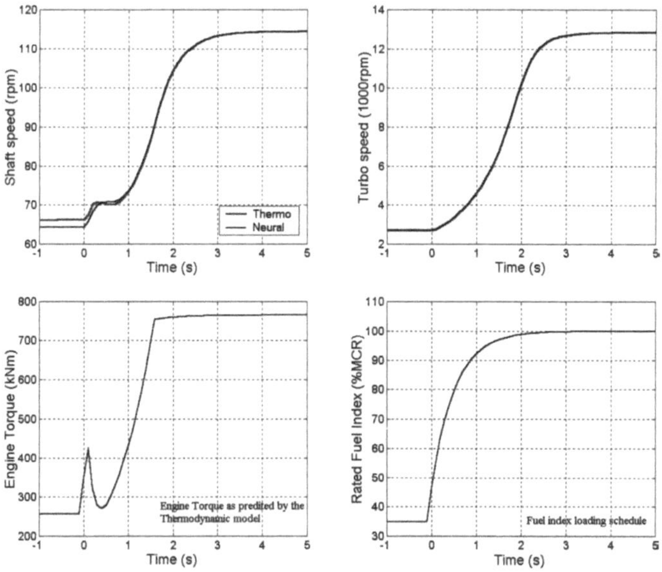

Figure 6.9 Effect of the filter on response with $\tau_f = 0.47$ s

In effect, use of the setpoint filter is allowing smoother engine running, without moderating the speed of response to a transient. More advanced filtering schemes can be used, *e.g.* with more than one pole, right-half *s*-plane zero(s) (non-minimum-phase filters), *etc.*, or even non-linear filters. Advanced filtering can be used for shaping the fuel index signal, so that a lower increase rate is observed at the beginning of the transient, where A/F mismatch mostly occurs, or for dealing with other operational problems.

6.3 Full-state-feedback Control of the Marine Plant

6.3.1 Theoretical Background

As was seen in Section 6.2, supervisory control can optimise plant operation and engine running, as well as increase the reliability of the installation. The use of non-linear optimisation theoretical results in combination with intelligent modelling and control, as well as the physical principles of thermodynamics, can be used for this purpose. However, a major prerequisite for the applicability of the above is that an appropriate feedback control scheme maintains plant operation as close as possible to its nominal status, which has been subject to extensive reliability analysis and optimisation. The type of control examined in this chapter towards this objective is full-state-feedback linear control, whilst the PID control law has been covered in Chapter 4.

First, the control problem for the case of the linear UPM is formulated. Given the state equations of UPM with disturbance:

$$\dot{\mathrm{x}} = \mathrm{A} \cdot \mathrm{x} + \mathrm{B} \cdot u + \mathrm{d},$$

as well as the nominal (without parametric uncertainty in the system matrices) perturbation model (NPM):

$$\dot{\mathrm{x}} = \mathrm{A}_0 \cdot \mathrm{x} + \mathrm{B}_0 \cdot u + \mathrm{d} \tag{6.16}$$

it is desired to determine the gain matrix K of the full-state-feedback control law

$$u = \mathrm{K} \cdot \mathrm{x} \tag{6.17}$$

so that:

$$|n| \leq \delta n \text{ and } |n_{TC}| \leq \delta n_{TC} \tag{6.18}$$

provided that both disturbance and uncertainties in the elements of the system matrices are bounded. The approach will be to establish a method for the design of a state-feedback controller with gain matrix K that provides adequate disturbance rejection for the NPM. Then, the application of standard theoretical, methodological tools will be examined for robustness assessment of the closed-loop system against the parametric uncertainty of the open-loop system matrices.

Although, robust control of linear systems is a subject that has been widely covered in control literature, the approach above bears some new features not usually encountered in standard theoretical methods. To explain the similarities and differences, a brief outline of the standard procedure of H∞ controller synthesis is given below. This synthesis method is based on the notion of the H∞-norm of a transfer function matrix; this, in turn, is based upon the singular values for a $(p \times m)$ real matrix M, which are the square roots of the purely positive eigenvalues of the non-negative-definite matrix $(\mathrm{M}^{\mathrm{T}} \cdot \mathrm{M})$ [55,56]. *I.e.* the set of the singular values of M is:

$$\left\{\sigma_i > 0 : \left(\mathrm{M}^{\mathrm{T}}\mathrm{M} - \sigma_i^2 \cdot \mathrm{I}_m\right) \cdot \vec{\mathrm{x}} = \vec{0}_m\right\} \tag{6.19}$$

It can be proven that the number of singular values is equal to the rank of M, $k_{RANK} \leq \min\{p, m\}$; moreover, singular value indexing, i, corresponds to their descending order, *i.e.*:

$$\sigma_1 \geq \sigma_2 \geq \ldots \geq \sigma_{k_{RANK}} > 0 \tag{6.20}$$

Therefore, the largest singular value of M, denoted $\sigma_{\max}(\mathrm{M})$, is equal to σ_1.

When M is the transfer function matrix of a linear system, *i.e.* a matrix function of complex frequency s, M(s), then the singular value $\sigma_{\max}(\mathrm{M}(s = j\omega))$ becomes a function of frequency ω. Then, the H∞-norm of transfer function matrix M(s), denoted $\|\mathrm{M}(s)\|_\infty$, is defined as:

$$\|\mathrm{M}(s)\|_\infty = \sup_{\omega \geq 0}\{\sigma_{\max}(\mathrm{M}(j\omega))\} \tag{6.21}$$

i.e. the supremum or least upper bound of the scalar, real-valued function $\sigma_{\max}(\mathrm{M}(j\omega))$.

In the case of linear systems the H∞-norm has a meaningful interpretation. Indeed, if the system setup in figure 6.10 is considered, and the linear system $\mathrm{M}(s)$ is excited with the following "sinusoidal" input vector u:

$$\mathrm{u}^{\mathrm{T}} = \mathrm{u}_0^{\mathrm{T}} \cdot \sin(\omega t) = \left[u_{01} \cdot \sin(\omega t) \quad u_{02} \cdot \sin(\omega t) \quad \ldots \quad u_{0m} \cdot \sin(\omega t) \right] \tag{6.22}$$

then the system output vector y will also contain sinusoidal functions of the same frequency ω and amplitudes $\mathrm{y}_0^{\mathrm{T}} = \left[y_{01} \quad y_{02} \quad \cdots \quad y_{0m} \right]$.

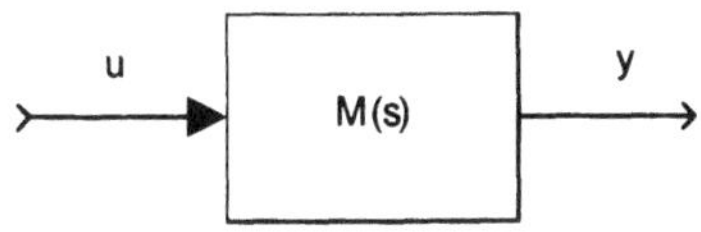

Figure 6.10 MIMO dynamical system with input–output description

Furthermore, the Euclidean norm of the output, $\|\mathrm{y}\|$:

$$\|\mathrm{y}\| = \|\mathrm{y}\|_2 = \sqrt{\sum_{i=1}^{p} y_{0i}^2} \tag{6.23}$$

will satisfy the following relationship:

$$\|\mathrm{y}\| \le \|\mathrm{M}(s)\|_\infty \cdot \|\mathrm{u}\| \tag{6.24}$$

where $\|\mathrm{u}\|$ is the Euclidean norm of the input vector.

The above theory is now applied to NPM with linear full-state-feedback controls. The closed-loop transfer function matrix $\mathrm{G}(s)$ from disturbance to the state vector is as follows:

$$\left. \begin{array}{l} \dot{\mathrm{x}} = \mathrm{A}_0 \cdot \mathrm{x} + \mathrm{B}_0 \cdot u + \mathrm{d} \\ u = \mathrm{K} \cdot \mathrm{x} \end{array} \right\} \Rightarrow \mathrm{x}(s) = \mathrm{G}(s) \cdot \mathrm{d}(s) = \left(s\mathrm{I}_2 - \mathrm{A}_0 - \mathrm{B}_0 \cdot \mathrm{K} \right)^{-1} \cdot \mathrm{d}(s) \tag{6.25}$$

In conclusion, one way to tackle the problem of propeller load disturbance rejection is to select the controller gain matrix K so that $\|\mathrm{G}(s)\|_\infty$ is below a specified value. In the literature this problem is referred to by the term suboptimal H∞ control problem. A solution, in terms of a full-state-feedback controller K, can be found provided that certain conditions concerning the structure of the plant, in comparison with the upper design bound for $\|\mathrm{G}(s)\|_\infty$, hold. The solution is basically a trial-and-error procedure, requiring repeated solutions of an associated mini–max differential game including an algebraic Riccati equation, until termination conditions are met [57].

Furthermore, the Structured Singular Value (SSV, μ) analysis and synthesis framework, based on the concept of Linear Fractional Transformation, allows dealing with problems of disturbance rejection in the case of uncertain plants [58–60]. Specifically, robust stability and performance (Main Loop Theorem) tests can be applied to closed-loop uncertain systems, providing validation of the designs. Moreover, the solution procedure for the H∞ control problem is generalised to the D–K iteration in order to finally provide a double iterative optimisation procedure, which, if terminated successfully, provides one with a controller ensuring both disturbance rejection and robustness of the closed-loop plant.

All the above theoretical background is covered in detail in the literature and could have been applied to the case of marine plants as well. However, a series of practical limitations, explained in Section 6.3.2, has motivated us towards a more application-specific approach, instead of just adopting an existing one. The resulting technique, except for bypassing the limitations below, proves to be effective not only in the case of marine propulsion plants, but also in the case of linear plants with a single (1D) source of disturbance and a state-space description available.

6.3.2 Practical H∞-norm Requirements

As already mentioned, control requirements for both the NPM and UPM are:

$$|n| \leq \delta n \text{ and } |n_{TC}| \leq \delta n_{TC} \tag{6.26}$$

In Chapter 5, it was argued that:

$$\mathrm{d}^{\mathrm{T}} = \begin{bmatrix} -I^{-1} \cdot \left(N_0 \cdot k_Q + \dot{N}_0 \cdot \Delta I\right) & 0 \end{bmatrix} = \begin{bmatrix} d_1 & 0 \end{bmatrix} \Rightarrow \|\mathrm{d}\| = |d_1| \tag{6.27}$$

By substituting, the closed-loop transfer function matrix of the NPM with linear state feedback controls becomes:

$$\begin{bmatrix} n(s) \\ n_{TC}(s) \end{bmatrix} = \frac{1}{p_c(s)} \cdot \begin{bmatrix} s - a_{22} - b_2 k_2 & a_{12} + b_1 k_2 \\ a_{21} + b_2 k_1 & s - a_{11} - b_1 k_1 \end{bmatrix} \cdot \begin{bmatrix} d_1(s) \\ 0 \end{bmatrix}$$

$$\Downarrow$$

$$\begin{bmatrix} n(s) \\ n_{TC}(s) \end{bmatrix} = \begin{bmatrix} G_1(s) \\ G_2(s) \end{bmatrix} \cdot d_1(s) = \begin{bmatrix} \dfrac{s - a_{22} - b_2 k_2}{p_c(s)} \\ \dfrac{a_{21} + b_2 k_1}{p_c(s)} \end{bmatrix} \cdot d_1(s) \tag{6.28}$$

where the closed-loop characteristic polynomial $p_c(s)$ in the above is given by:

$$\begin{aligned} p_c(s) = s^2 - (a_{11} + a_{22} + b_1 k_1 + b_2 k_2) \cdot s + (a_{11} a_{22} - a_{12} a_{21}) \\ + (a_{22} b_1 - a_{12} b_2) k_1 + (a_{11} b_2 - a_{21} b_1) k_2 \end{aligned} \tag{6.29}$$

Note that, in the above, index 0 for the elements of matrices $\mathrm{A}_0 = [a_{ij,0}]$ and $\mathrm{B}_0 = [b_{i,0}]$, indicating the nominal values of the elements of UPM system matrices A and B, has been temporarily dropped in order to simplify the formulae. Later, however, when robustness of the closed-loop system is examined, index 0 is restored.

Applying the results of Section 6.3.1 for the H∞-norm in the scalar case, the following inequalities are obtained for $|n|$ and $|n_{TC}|$:

$$|n| \le \|G_1(s)\|_\infty \cdot |d_1| \text{ and } |n_{TC}| \le \|G_2(s)\|_\infty \cdot |d_1| \tag{6.30}$$

In the above, the H∞-norm calculations include only scalar (closed-loop) transfer functions and not a transfer function matrix. Furthermore, from the analysis in Section 5.4.2 it can be readily concluded that disturbance d_1 can be bounded in an interval symmetrical around zero, yielding:

$$|d_1| \le \delta d \tag{6.31}$$

Therefore, by combining the above:

$$|n| \le \|G_1(s)\|_\infty \cdot \delta d \text{ and } |n_{TC}| \le \|G_2(s)\|_\infty \cdot \delta d \tag{6.32}$$

Finally, the requirements for $|n|$ and $|n_{TC}|$ are translated to the following H∞-norm requirements (which will be used in design) for the scalar transfer functions $G_1(s)$ and $G_2(s)$ from disturbance d_1 to shaft and turbo rpm, n and n_{TC} respectively:

$$\|G_1(s)\|_\infty \le \frac{\delta n}{\delta d} = G_{10} \text{ and } \|G_2(s)\|_\infty \le \frac{\delta n_{TC}}{\delta d} = G_{20} \tag{6.33}$$

It is now argued that a "formal" requirement for $\|\mathrm{G}(s)\|_\infty$, which can guarantee disturbance rejection individually for each one of the marine plant state variables equivalent to the rejection provided by the two "practical" H∞-norm requirements above, is the following:

$$\|\mathrm{G}(s)\|_\infty \le G_{0\min} = \min\{G_{10}, G_{20}\} \tag{6.34}$$

Indeed:

$$\|\mathrm{x}\| \le \|\mathrm{G}(s)\|_\infty \cdot \|\mathrm{d}\| \le G_{0\min} \cdot \|\mathrm{d}\| \Rightarrow \left(|n|^2 + |n_{TC}|^2\right)^{1/2} \le G_{0\min} \cdot |d_1| \tag{6.35}$$

Thus:

$$\begin{aligned} |n| &\le \sqrt{|n|^2 + |n_{TC}|^2} \le G_{0\min} \cdot |d_1| \le G_{10} \cdot |d_1| \\ |n_{TC}| &\le \sqrt{|n|^2 + |n_{TC}|^2} \le G_{0\min} \cdot |d_1| \le G_{20} \cdot |d_1| \end{aligned} \tag{6.36}$$

However, such a specification for $\|\mathrm{G}(s)\|_\infty$ can be expected to be harder to meet, as mentioned in literature, *e.g.* [44].

On the other hand, if the upper bound chosen for $\|\mathrm{G}(s)\|_\infty$ is larger than $G_{0\min}$ then it is not guaranteed that the two separate practical requirements for $G_1(s)$ and $G_2(s)$ are met. The argumentation starts by considering a bound:

$$\|\mathrm{G}(s)\|_\infty^2 \le G_{0\min}^2 + \varepsilon_0 \text{ where } \varepsilon_0 > 0 \tag{6.37}$$

It is assumed without loss of generality that $G_{0\min} = G_{10}$. The above requirement can be met by the following combination of inequalities holding for the Euclidean norm of each one of the state variables individually:

$$\begin{aligned} |n|^2 &\le \left(G_{10}^2 + \varepsilon\right)\cdot |d_1|^2 \\ |n_{TC}|^2 &\le (\varepsilon_0 - \varepsilon)\cdot |d_1|^2 \end{aligned} \tag{6.38}$$

for some ε in the interval $0 < \varepsilon < \varepsilon_0$. By combining the above, it is easily concluded that the requirement for $\|\mathrm{G}(s)\|_\infty$ is met; however, it cannot be guaranteed that $|n| \le \delta n$ or, equivalently, $\|G_1(s)\|_\infty \le G_{10}$.

The above conclusions are now exemplified in the case of the open-loop marine plant, whose transfer functions were obtained in Section 5.5.1. No control is assumed, which is translated to $f_R \equiv 0$. Therefore:

$$\begin{bmatrix} n(s) \\ n_{TC}(s) \end{bmatrix} = \begin{bmatrix} T_{\mathrm{xd},1}(s) \\ T_{\mathrm{xd},2}(s) \end{bmatrix} \cdot d_1(s) \tag{6.39}$$

Both $T_{\mathrm{xd},1}(s)$ and $T_{\mathrm{xd},2}(s)$ are all-pole transfer functions with real poles. Thus:

$$\|T_{\mathrm{xd},1}(s)\|_\infty = |T_{\mathrm{xd},1}(j0)| \text{ and } \|T_{\mathrm{xd},2}(s)\|_\infty = |T_{\mathrm{xd},2}(j0)| \tag{6.40}$$

As analysed in Section 5.5.1, $T_{\mathrm{xd},1}(s)$ has a single pole at $s = a_{11}$; $T_{\mathrm{xd},2}(s)$ has two poles located at $s = a_{11}$ and $s = a_{22}$. Note that in the case of $T_{\mathrm{xd},1}(s)$ zero-pole cancellation occurs for $s = a_{22}$ and this pole becomes unobservable. In effect:

$$\|T_{\mathrm{xd},1}(s)\|_\infty = \frac{1}{|a_{11}|} \approx \frac{I}{2K_{Q0}N_0} \text{ and } \|T_{\mathrm{xd},2}(s)\|_\infty = \left|\frac{a_{21}}{a_{11}a_{22}}\right| = \frac{I}{2K_{Q0}N_0}\cdot\frac{|q_{TC,1}|}{|q_{TC,2}|} \tag{6.41}$$

Taking into account typical numerical values for the physical parameters of common marine plants it can be supported that if the powerplant is left without feedback control then prohibitive overspeed, especially in the near-MCR operating range, can be expected to occur for shaft rpm. Indeed, as seen in Section 5.4.3 for the MAN-B&W 6L60MC propulsion powerplant, the following holds:

$$|q_{TC,1}| \ll |q_{TC,2}| \Rightarrow \|T_{\mathrm{xd},2}(s)\|_\infty \ll \|T_{\mathrm{xd},1}(s)\|_\infty \text{ and } |n_{TC}| \ll \delta n_{TC} \tag{6.42}$$

Furthermore:

$$N \ll N_{TC} \Rightarrow \delta n \ll \delta n_{TC} \tag{6.43}$$

Taking into account the above, it is possible that, although the following requirement is readily met, even without the need for feedback control, engine overspeed still occurs (in terms of shaft rpm).

$$\frac{\sqrt{|n|^2 + |n_{TC}|^2}}{|d_1|} \le \|\mathrm{G}(s)\|_\infty \le \frac{\sqrt{(\delta n)^2 + (\delta n_{TC})^2}}{\delta d} \tag{6.44}$$

Rating the values of n and n_{TC} with respect to N_0 and N_{TC0}, or the MCR values, will improve the situation only slightly, as turbocharger speed has a rather small dependence on propeller torque coefficient, as demonstrated by the magnitude of $\|T_{\mathrm{xd},2}(s)\|_\infty$. Weighting the contribution of n and n_{TC} in $\|\mathrm{x}\|$ will also not have a significant effect if the contributions of δn and δn_{TC} in the upper bound are weighted in accordance. Therefore, it is required to apply repeated reductions in the upper bound of $\|G(s)\|_\infty$ and redesign the H∞ controller so that the new requirement is met. Then, the requirement for $|n|$ should be checked separately; if met, the generated controller is acceptable; otherwise, a further reduction in the $\|G(s)\|_\infty$ requirement is needed until value $(\delta n/\delta d)$ is reached.

In most cases the above procedure is expected to work out after the bound for $\|G(s)\|_\infty$ has been reduced to a value closer to $G_{0\min}$. However, a simpler approach would be to design the controller so that the individual H∞-norm requirements for $G_1(s)$ and $G_2(s)$ are satisfied. Moreover, when standard H∞ synthesis procedures, which rely greatly on numerical iterative procedures, are used, the poles and zeros of the closed-loop system are "arbitrarily" placed. In effect, robustness against unmodelled plant dynamics is expected to be harder both to achieve and check. In effect, a technique applicable to second-order systems with scalar disturbance is given below. The proposed technique has a significantly reduced numerical part and allows for deeper insight to be gained, as it is essentially a pole placement method.

6.3.3 Marine Plant Regulator Synthesis

Having analysed the major aspects of marine plant dynamics and the problems of direct application of the H∞ controller synthesis method, an alternative design procedure for the marine plant case is presented. The following assumption, valid for marine plants, was introduced in Section 5.5.1:

$$q_{E,2} \approx 0 \Rightarrow a_{12} = \frac{q_{E,2}}{J} \approx 0 \tag{6.45}$$

This assumption is justified by the fact that the turbocharging system is appropriately matched to the engine so that adequate combustion air is driven to the engine cylinders. Therefore, it can be assumed that engine torque is practically insensitive to turbocharger rpm variations for an adequately wide range of values. Under this assumption the closed-loop characteristic polynomial $p_c(s)$ obtained previously becomes:

$$p_c(s) = s^2 - (a_{11} + a_{22} + b_1k_1 + b_2k_2)\cdot s + a_{11}a_{22} + a_{22}b_1k_1 + (a_{11}b_2 - a_{21}b_1)k_2 \tag{6.46}$$

For design purposes the following standard form for the scalar transfer functions of interest, $G_1(s)$ and $G_2(s)$, will be used:

$$G_1(s) = \frac{s+z}{(s+p_1)\cdot(s+p_2)}, \quad G_2(s) = \frac{a_{21}+b_2k_1}{(s+p_1)\cdot(s+p_2)} \tag{6.47}$$

where:

$$z = -a_{22} - b_2 k_2 \text{ and } p_c(s) = (s + p_1)\cdot(s + p_2) \tag{6.48}$$

By equating the coefficients of the two forms of $p_c(s)$, the following two equations for the regulator gains k_1 and k_2 are obtained:

$$\begin{aligned} b_1 k_1 + b_2 k_2 &= -(a_{11} + a_{22}) - (p_1 + p_2) \\ a_{22} b_1 k_1 + (a_{11} b_2 - a_{21} b_1) k_2 &= p_1 p_2 - a_{11} a_{22} \end{aligned} \tag{6.49}$$

The values of system poles and zero can be determined by the requirements for $\|G_1(s)\|_\infty$ and $\|G_2(s)\|_\infty$. The peak of the magnitude Bode plot of each one of the transfer functions, however, depends on its shape. In the following, the closed-loop poles and zero will be assumed to be purely real and stable. Then, for a $G_2(s)$ that has solely two poles the peak value of the Bode plot can be calculated as the value of the transfer function at $\omega = 0$, *i.e.*:

$$\|G_2(s)\|_\infty = |G_2(j0)| = \frac{|a_{21} + b_2 k_1|}{p_1 \cdot p_2} \tag{6.50}$$

For $G_1(s)$, the peak value and the frequency at which it occurs depend upon the relative position of closed-loop zero z with respect to the smallest of the poles. Specifically:

$$\|G_1(s)\|_\infty = \begin{cases} |G_1(jp_1)| = \sqrt{\dfrac{z^2 + p_1^2}{2p_1^2(p_1^2 + p_2^2)}}, & \text{if } z < p_1 \\ |G_1(j0)| = \dfrac{z}{p_1 \cdot p_2}, & \text{if } z \geq p_1 \end{cases} \tag{6.51}$$

where, without loss of generality, it has been assumed that $p_1 = \min\{p_1, p_2\}$. By assuming the second expression (as it is easier to manipulate), the following constraint inequalities for the location of the real poles and zero are deduced:

$$\begin{aligned} &0 < p_1 \leq z = -a_{22} - b_2 k_2 \leq p_1 p_2 \cdot G_{10} \\ &\text{and } |a_{21} + b_2 k_1| \leq p_1 p_2 \cdot G_{20} \end{aligned} \tag{6.52}$$

where G_{10} and G_{20} are the required upper bounds for $\|G_1(s)\|_\infty$ and $\|G_2(s)\|_\infty$ respectively.

The complete H∞ admissible value set for p_1, p_2 can be obtained by re-formulating the inequalities above on the basis of a convenient change of variables. Indeed, it can be easily seen that, if the following variables are used (instead of p_1, p_2), the majority of the inequalities to be satisfied become linear:

$$S = p_1 + p_2,\ P = p_1 p_2 \tag{6.53}$$

Controller gains can be calculated on the basis of S and P, from the pole-placement equations, as follows:

$$\begin{bmatrix} k_1 \\ k_2 \end{bmatrix} = \begin{bmatrix} b_1 & b_2 \\ a_{22}b_1 & a_{11}b_2 - a_{21}b_1 \end{bmatrix}^{-1} \cdot \begin{bmatrix} -(a_{11}+a_{22})-S \\ P - a_{11}a_{22} \end{bmatrix}$$
$$\Downarrow \tag{6.54}$$
$$\begin{Bmatrix} k_1 = k_{1\varnothing} + k_{1S} \cdot S + k_{1P} \cdot P \\ k_2 = k_{2\varnothing} + k_{2S} \cdot S + k_{2P} \cdot P \end{Bmatrix}$$

where $k_{i\varnothing}, k_{iS}, k_{iP}, i=1,2$ are readily calculated using the values of the open-loop system parameters (elements of the open-loop system matrices A and B).

The inequalities to be satisfied are therefore formulated straightforwardly for *P* and *S* from the H∞-norm requirements presented previously. Specifically, the following five linear inequalities can finally be obtained for pair (*S,P*):

$$\begin{aligned} &0 < -a_{22} - b_2 \cdot (k_{2\varnothing} + k_{2S} \cdot S + k_{2P} \cdot P) \leq G_{10} \cdot P \\ &\left| a_{21} + b_2 \cdot (k_{1\varnothing} + k_{1S} \cdot S + k_{1P} \cdot P) \right| \leq G_{20} \cdot P \end{aligned} \tag{6.55}$$

Note that one of the inequalities, namely $p_1 \leq z$, has been reduced to $0 < z$, *i.e.* the requirement for $G_1(s)$ to be minimum-phase. If $p_1 \leq z$ was not omitted, however, the system of inequalities could not be obtained in the above linear form, a form that is much easier to solve either analytically or numerically.

Therefore, a value set for *S, P* will be determined, which in turn can provide the value set for the closed-loop poles p_1, p_2. These are determined (for each value pair (*S,P*)) as roots of the following second-order parametric equation:

$$\chi^2 + S \cdot \chi + P = 0 \tag{6.56}$$

provided that the following constraints are satisfied:

- $S > 0$, so that the closed-loop system is stable,
- $S^2 \geq 4P$, so that p_1, p_2 are real, and, finally,
- $p_1 = \min\{p_1, p_2\} \leq z = -a_{22} - b_2 k_2$ in order for the calculated value of $\|G_1(s)\|_\infty$ to be valid.

In conclusion, the admissible subset of $\mathbb{R}^2$ for the dyad (p_1, p_2) will be determined, as well as the corresponding value of closed-loop zero *z*. The final choice of the closed-loop poles and zero of NPM, through controller gains, can be done from this set in order to fulfil other concerns, such as robustness against UPM modelling uncertainty. This is exemplified in the following test case.

6.3.4 Test Case: MAN B&W 6L60MC Marine Plant

For testing the controller synthesis procedure presented above, a propulsion plant equipped with the MAN B&W 6L60MC marine Diesel engine directly coupled to the propeller has been examined. The values of the system parameters, such as partial torque derivatives, required for the calculation of the transition matrix A and matrix B of the state equation are given in Table 6.1.

Table 6.1 Specifics of powerplant with MAN B&W 6L60MC engine

$q_{E,1} = 0.46$ N m/rpm	$q_{TC,1} = 4.70$ N m/rpm
$q_{E,2} = 0.0$ N m/rpm	$q_{TC,2} = 180$ N m/rpm
$q_{E,0} = 9.0$ kN m/%	$q_{TC,0} = 700$ N m/%
$N_0 = 114.6$ rpm	$N_{TC0} = 12840$ rpm
$I = 59800$ kg m^2	$I_{TC} = 4.83$ kg m^2
$K_{Q0} = 0.0578$ kN m/rpm^2	

By substituting the values in Table 6.1, the following matrices A, B are obtained

$$\mathrm{A} = \begin{bmatrix} -2.1319 & 0.0 \\ 9.2923 & -355.8744 \end{bmatrix} \mathrm{s}^{-1}, \quad \mathrm{B} = \begin{bmatrix} 1.4372 \\ 1384 \end{bmatrix} \frac{\text{rpm/s}}{\%\text{index}}$$

The linear state-space NPM is shown in Figure 6.11.

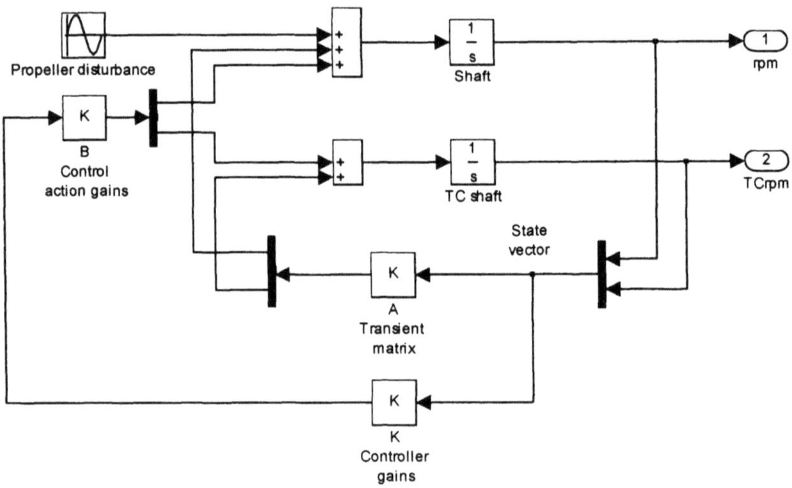

Figure 6.11 Linear perturbation model of marine plant for regulator synthesis

In order to estimate the required bounds for $\|G_1(s)\|_\infty$ and $\|G_2(s)\|_\infty$ the maximum absolute value of uncertainty δk_Q is assumed to be 10% of the nominal value K_{Q0}, *i.e.* approximately 6 N m/rpm^2. The introduced disturbance d_1 is therefore expected to lie in the range:

$$|d_1| = \frac{N_0^2 \cdot |k_Q|}{I} \leq \delta d = 12.7 \text{ rpm/s}$$

A fluctuation no larger than 1.2 rpm for shaft rpm and 30 rpm for turbo rpm are the design targets. Therefore, the upper bounds for the H∞-norm of the two transfer functions of interest are:

$$G_{10} = \frac{1.2 \text{ rpm}}{12.7 \text{ rpm/s}} = 0.0945 \text{ s or -20.5 dB}$$

$$G_{20} = \frac{30 \text{ rpm}}{12.7 \text{ rpm/s}} = 2.3622 \text{ s or 7.5 dB}$$

The open-loop poles (K=0) are the eigenvalues of the transient matrix A, *i.e.* –2.319 and –355.8744. The values for $\|G_1(s)\|_\infty$ and $\|G_2(s)\|_\infty$ when no feedback is applied are:

$$\|G_1(s)\|_\infty = -6.6 \text{ dB and } \|G_2(s)\|_\infty = -38.3 \text{ dB}$$

Although the requirement for $\|G_2(s)\|_\infty$ is met by far, the upper bound for $\|G_1(s)\|_\infty$ is exceeded by approximately 14.0 dB. The magnitude and phase Bode plots of the open-loop plant are shown in Figure 6.12. Evidently, redistribution of the effect of the disturbance is sought after and, as demonstrated, can be achieved by the use of an appropriate linear state-feedback linear control law. As long as the values of $\|G_1(s)\|_\infty$ and $\|G_2(s)\|_\infty$ are below the required bounds the poles of the closed-loop can be placed arbitrarily, so that other requirements, such as robustness against neglected dynamics, are met.

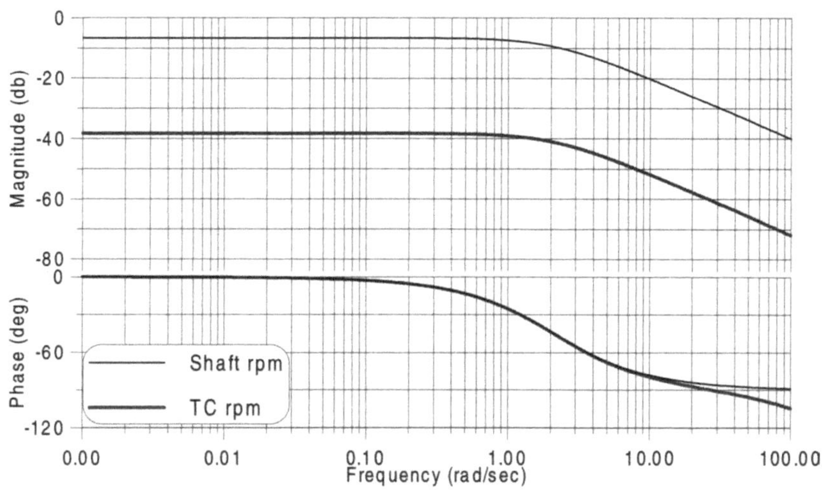

Figure 6.12 Bode plots of open-loop transfer functions from disturbance to states

Moreover, the open-loop system is clearly unobservable. The unobservable pole is the fast pole $p_2 = a_{22}$ that coincides with the open-loop zero z. This fact can be interpreted physically by the insensitivity of shaft rpm to turbo rpm variations, provided that the turbo maintains adequate air flow rate for complete combustion and adequate scavenging in the engine cylinders. This is normally valid for marine engines, as typical (steady-state) air-to-fuel ratio values are 30.0–40.0, whilst the perfect combustion limit is less than 20.0.

In this example the following two cases are included.

(i) Double real pole at $-p_1 = -p_2 = -19.9$. This choice can be achieved by setting the state feedback gains to:

$$K = [k_1 \quad k_2] = [-0.6270 \quad 0.2306] \tag{6.57}$$

The zero of $G_1(s)$ is then

$$z = 36.8 > 19.9 = p_{1,2} \tag{6.58}$$

The values achieved for the H∞-norm of the two transfer functions of interest are:

$$\|G_1(s)\|_\infty = -20.65 \text{ dB and } \|G_2(s)\|_\infty = 6.72 \text{ dB}$$

Compliance with the H∞ requirements is therefore concluded. The magnitude and phase Bode plots of the compensated system in this case are shown in Figure 6.13.

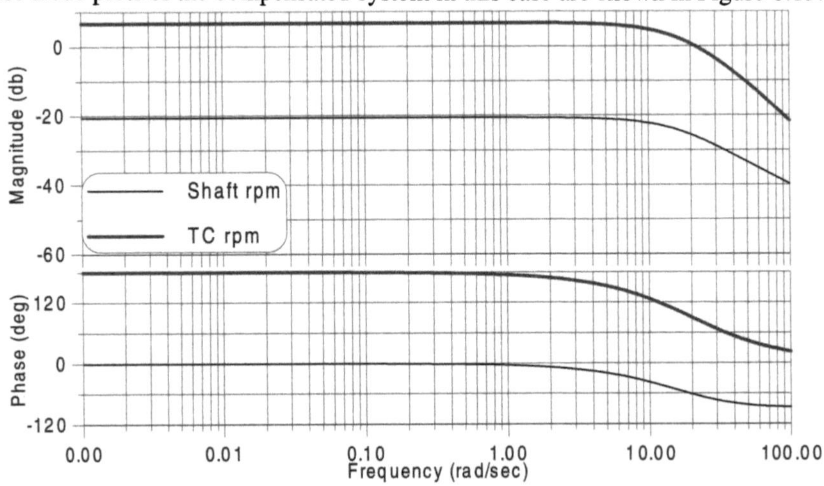

Figure 6.13 Bode plots of the closed-loop transfer functions – case (i)

(ii) Discrete real poles at $-p_1 = -11.0$ and $-p_2 = -250.0$. This choice can be achieved by setting the state feedback gains to:

$$K = [k_1 \quad k_2] = [-4.3256 \quad 0.0746] \tag{6.59}$$

The zero of $G_1(s)$ is then

$$z = 252.7 > 11.0 = p_1 \tag{6.60}$$

The values achieved for the H∞-norm of the two transfer functions of interest are:

$$\|G_1(s)\|_\infty = -20.74 \text{ dB and } \|G_2(s)\|_\infty = 6.74 \text{ dB}$$

Compliance with the H∞ requirements is therefore concluded, for this choice, as well. The magnitude and phase Bode plots of the compensated system in this case are shown in Figure 6.14.

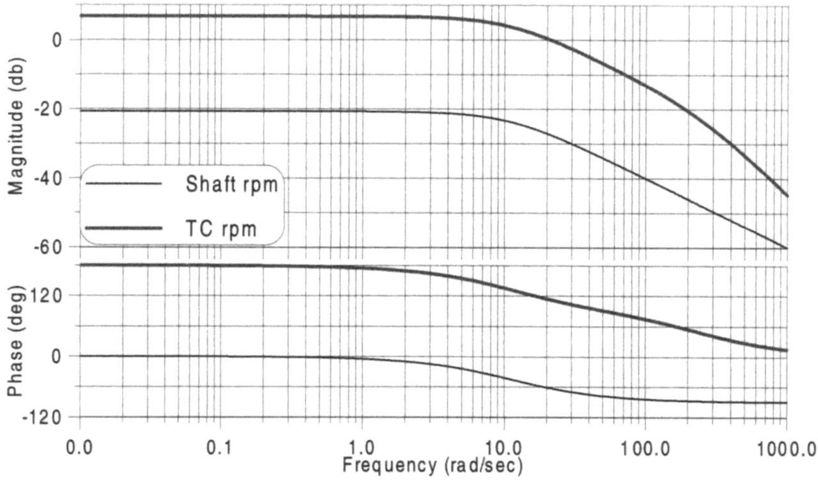

Figure 6.14 Bode plots of the closed-loop transfer functions – case (ii)

Also, it can be easily seen that the transfer function from d_1 to the control action y can be made minimum phase with appropriate selection of the controller gains. Furthermore, choice of the controller can be based upon any other additional requirements for the closed-loop system transfer function matrix. For example, the double real pole of case (i) may be too close to neglected fast system poles, thus compromising robustness of the closed-loop system against unmodelled fast dynamics. On the other hand, a slower pole p_1 is achieved in case (ii), providing enhanced robustness against fast dynamics. However, zero-pole cancellation may occur, as the fast system pole p_2 is located close to the zero z of $G_1(s)$. This can make the system unobservable, as in the open-loop case, if some parametric uncertainty is introduced.

Finally, note that the effect of limiting functions, such as scavenging pressure limiter, is essentially substituted by the turbo shaft rpm feedback. Indeed, this is the case for both the controllers examined above $k_2 > 0$. Therefore, the fuel index increase when the turbo rpm is below the required value is moderated, in order to prevent a potential lack in air charge.

6.3.5 Robustness Against Model Uncertainty

As already demonstrated, the UPM system matrices incorporate significant uncertainty in their elements. Having obtained a controller with the synthesis procedure above, the resulting closed-loop system has to be checked for robustness. Two major forms of robustness can be clearly distinguished:

- robust stability of the closed-loop system, and
- robust performance of the closed-loop system.

For each one of the above there exists a major theoretical result (test) for a given linear system. The related results are as follows:

- Robust stability of linear systems – **Theorem of Kharitonov**

This result allows guaranteeing the stability of a family of characteristic polynomials whose coefficients are not fixed, but lie in specified intervals (interval polynomials), by checking if four well-defined "edge" polynomials of the family are stable.

- Robust performance of linear systems – **Zero-Exclusion Principle**

This result allows guaranteeing that the H∞-norm of a family of proper scalar transfer functions, whose coefficients are not fixed, but lie in specified subsets of $\mathbb{R}$, remains below a specified performance level by checking robust stability of an appropriately derived complex polynomial. Robust stability of complex polynomials with uncertain parameters can be checked, in turn, using the Zero Exclusion Condition.

A full statement of the above two theoretical results is rather lengthy and can be found easily in the literature with different versions of their proof [61–71]; therefore it is omitted here. However, a common requirement for the application of both of the above results is to clearly define the subsets of $\mathbb{R}$ where parametric uncertainty of the transfer functions lies. Specifically, for marine plants, the following two nominal closed-loop transfer functions have been obtained after the introduction of full-state-feedback controls to the NPM:

$$\begin{bmatrix} n(s) \\ n_{TC}(s) \end{bmatrix} = \begin{bmatrix} G_{1,\text{nom}}(s) \\ G_{2,\text{nom}}(s) \end{bmatrix} \cdot d_1(s) = \begin{bmatrix} \dfrac{s - a_{22,0} - b_{2,0}k_2}{p_{c,0}(s)} \\ \dfrac{a_{21,0} + b_{2,0}k_1}{p_{c,0}(s)} \end{bmatrix} \cdot d_1(s) \tag{6.61}$$

where the nominal closed-loop characteristic polynomial $p_{c,0}(s)$ in the above is given by:

$$\begin{aligned} p_{c,0}(s) = s^2 - (a_{11,0} + a_{22,0} + b_{1,0}k_1 + b_{2,0}k_2) \cdot s + \ (a_{11,0}a_{22,0} - a_{12,0}a_{21,0}) \\ + (a_{22,0}b_{1,0} - a_{12,0}b_{2,0})k_1 + (a_{11,0}b_{2,0} - a_{21,0}b_{1,0})k_2 \end{aligned} \tag{6.62}$$

Note that in the above expressions the notation for the elements of the open-loop UPM has been fully restored, as robustness considerations are examined now. One is reminded that the value assigned to a_{12} during synthesis was zero, for the reasons explained previously. However, the full form of the characteristic polynomial is required for robustness analysis. Finally, $G_{1,\text{nom}}(s)$ and $G_{2,\text{nom}}(s)$ denote the fixed-coefficients scalar transfer functions obtained when the nominal values of the elements of the UPM system matrices are substituted in $G_1(s)$ and $G_2(s)$.

In the literature, such as [71], uncertainty structure of linear, scalar transfer functions is distinguished according to the form of the functional dependence, which connects the numerator and denominator polynomial coefficients with the uncertain parameters. In the case of marine plants, the following standard form is considered for $G_1(s)$ and $G_2(s)$:

$$G_1(s) = \frac{s + c_{\mathrm{N1}}}{s^2 + c_{\mathrm{1D}} \cdot s + c_{\mathrm{0D}}} \text{ and } G_2(s) = \frac{c_{\mathrm{N2}}}{s^2 + c_{\mathrm{1D}} \cdot s + c_{\mathrm{0D}}} \tag{6.63}$$

Obviously, the coefficients of the above polynomials are calculated using the elements of the UPM state matrices $\mathrm{A}=[a_{ij}]$ and $\mathrm{B}=[b_i]$. Therefore, they are not constant; they are functions of the following vector of state-space parametric uncertainty:

$$\rho^{\mathrm{T}} = [\Delta a_{11} \quad \Delta a_{12} \quad \Delta a_{21} \quad \Delta a_{22} \quad \Delta b_1 \quad \Delta b_2] \tag{6.64}$$

For example:

$$\begin{aligned} c_{\mathrm{1D}}(\rho) &= c_{\mathrm{1D}}(\Delta a_{11}, \Delta a_{22}, \Delta b_1, \Delta b_2) = a_{11} + a_{22} + b_1 k_1 + b_2 k_2 \\ &= (a_{11,0} + a_{22,0} + k_1 \cdot b_{1,0} + k_2 \cdot b_{2,0}) + (\Delta a_{11} + \Delta a_{22} + k_1 \cdot \Delta b_1 + k_2 \cdot \Delta b_2) \end{aligned} \tag{6.65}$$

Note that ρ belongs to a sphere of $\mathbb{R}^5$ with radius:

$$\|\rho\|^2 \leq (\delta a_{11})^2 + (\delta a_{12})^2 + (\delta a_{21})^2 + (\delta a_{22})^2 + (\delta b_1)^2 + (\delta b_2)^2 \tag{6.66}$$

In the above, as already mentioned, lower case delta denotes the uncertainty radius of elements of UPM system matrices A and B, which have been determined in detail in Section 5.4.2.

In effect, the coefficients of the transfer functions also lie in well-defined intervals; the bounds of these intervals have to be determined in order to assess closed-loop system robustness. This can be effectuated in a manner similar to the one used in Section 5.4.2 for handling uncertainty of marine plant physical parameters in order to obtain state-space parametric uncertainty. For example, in the case of c_{1D}:

$$\underline{c}_{\mathrm{1D}} \leq c_{\mathrm{1D}} \leq \overline{c}_{\mathrm{1D}} \tag{6.67}$$

where:

$$\begin{aligned} \underline{c}_{\mathrm{1D}} &= (a_{11,0} + a_{22,0} + k_1 \cdot b_{1,0} + k_2 \cdot b_{2,0}) - (\delta a_{11} + \delta a_{22} - k_1 \cdot \delta b_1 + k_2 \cdot \delta b_2) \\ \overline{c}_{\mathrm{1D}} &= (a_{11,0} + a_{22,0} + k_1 \cdot b_{1,0} + k_2 \cdot b_{2,0}) + (\delta a_{11} + \delta a_{22} - k_1 \cdot \delta b_1 + k_2 \cdot \delta b_2) \end{aligned} \tag{6.68}$$

In the above it has been assumed that $k_1 < 0$ and $k_2 > 0$, which is valid for common marine plants for the reasons explained in Section 6.3.4. Finally, for c_{0D} the determination of bounds is more complicated, as its expression includes products between different elements of UPM system matrices A and B; however, it is completely tractable by using the technique of Section 5.4.2.

If closely examined, independent bounding of all coefficients of closed-loop transfer functions $G_1(s)$ and $G_2(s)$ can be easily proven to be rather conservative, *i.e.* including many unrealistic cases. This can lead to rather conservative conclusions, too; for example, it can force one to reject a controller, because instability occurs for a combination of coefficients of $G_1(s)$ and $G_2(s)$ that cannot be achieved actually by any value inside the sphere specified for vector ρ.

One way to deal with this problem is to "tighten" the coverage of the uncertainty included in the coefficients of closed-loop transfer functions $G_1(s)$ and $G_2(s)$. From the expressions for these coefficients it can be easily seen that all functions are affine with respect to the elements of ρ; the only exception is $c_{0D}(\rho)$, which is, however, multiaffine. A real-valued function of several variables $F(\rho), \rho \in \mathbb{R}^l$, is said to be affine if it is the sum of a linear function plus a constant, *i.e.* if it has the following form:

$$F(\rho) = \vec{C}_C \cdot \rho + C_{C0} \tag{6.69}$$

where $\vec{C}_C^T = [C_{C1} \quad C_{C2} \quad \dots \quad C_{Cl}]$ and C_{C0} are constant. Moreover, a function $F(\rho), \rho \in \mathbb{R}^l$, is said to be multiaffine if the following condition holds for each component of ρ separately: provided that all components of ρ are fixed but one, the function is affine. Therefore, products between different elements of ρ are allowed, but not powers.

Extensive results concern the application of the Zero Exclusion Condition for polynomial stability to polynomials with coefficients having affine or multiaffine uncertainty structure, as in the case of marine plants. The Zero Exclusion Condition states that a polynomial $p(s,\rho), \rho \in S \subset \mathbb{R}^l$, is robustly stable, *i.e.* stable for all $\rho \in S$,, if:

- it is stable for some $\rho_\varnothing \in S$, and
- $p(j\omega,\rho) \neq 0, \forall \rho \in S, \forall \omega \in \mathbb{R}$.

As already mentioned, the above condition for polynomial stability is extended to the Zero Exclusion Principle, enabling one to verify robust performance of a closed-loop plant. In effect, a technique to parameterise uncertainty in the closed-loop transfer functions of a second-order system with scalar disturbance on the basis of open-loop state-space uncertainty was given. This allows the application of standard robustness assessment tools to the case of marine plants and other systems with similar dynamics.

6.4 State-feedback and Integral Control of the Marine Plant

6.4.1 Steady-state Error Analysis

A major problem with all feedback controller types that incorporate only error-proportional action is steady-state error. For example, as analysed in Section 4.1, P and PD controllers cannot completely eliminate steady-state error; however, it can be reduced by appropriately increasing the proportional gain.

In the case of feedback control the situation is not much different. As the gains of the linear, state-feedback control law are selected on the basis of transient response to propeller torque coefficient disturbance, the value of steady-state error is significant. Specifically, by using the Final Value Theorem of the Laplace transform the following is obtained for a step change in disturbance d_1:

$$\mathrm{x}(t \to \infty) = \lim_{s \to 0}\left(s \cdot \mathrm{x}(s)\right) = \mathrm{G}(0) \cdot d_1(t \to \infty)$$

$$\Downarrow$$

$$\begin{bmatrix} n(t \to \infty) \\ n_{TC}(t \to \infty) \end{bmatrix} = \begin{bmatrix} G_1(0) \\ G_2(0) \end{bmatrix} \cdot d_1(t \to \infty) = \begin{bmatrix} \dfrac{-a_{22} - b_2 k_2}{p_c(0)} \\ \dfrac{a_{21} + b_2 k_1}{p_c(0)} \end{bmatrix} \cdot d_1(t \to \infty) \tag{6.70}$$

In the above:

$$p_c(0) = (a_{11}a_{22} - a_{12}a_{21}) + (a_{22}b_1 - a_{12}b_2)k_1 + (a_{11}b_2 - a_{21}b_1)k_2 \tag{6.71}$$

If the assumption $a_{12} \approx 0$ is valid for steady-state then:

$$p_c(0) = p_{c,0}(0) = (a_{11} + b_1 k_1)a_{22} + (a_{11}b_2 - a_{21}b_1)k_2 \tag{6.72}$$

Furthermore, because as explained in Section 5.4.3:

$$d_1 = -\frac{N_0^2}{I} \cdot k_Q - \frac{\dot{N}_0}{I} \cdot \Delta I$$

and because in steady-state either $\dot{N}_0 = 0$ or $\Delta I = 0$ or both:

$$d_1(t \to \infty) = -\frac{N_0^2}{I} \cdot k_Q(t \to \infty) \tag{6.73}$$

Therefore, the following final expression is obtained for shaft and turbo rpm:

$$\begin{bmatrix} n(t \to \infty) \\ n_{TC}(t \to \infty) \end{bmatrix} = \begin{bmatrix} -a_{22} - b_2 k_2 \\ a_{21} + b_2 k_1 \end{bmatrix} \cdot \frac{N_0^2 \cdot I^{-1}}{(a_{11} + b_1 k_1)a_{22} + (a_{11}b_2 - a_{21}b_1)k_2} \cdot \left(-k_Q(t \to \infty)\right)$$

$$\Downarrow$$

$$\begin{bmatrix} n(t \to \infty) \\ n_{TC}(t \to \infty) \end{bmatrix} =$$

$$\begin{bmatrix} \dfrac{N_0^2 \cdot \left(q_{TC,2} + k_2 q_{TC,0}\right) \cdot k_Q\left(t \to \infty\right)}{\left(q_{E,1} - 2K_{Q0}N_0 + q_{E,0}k_1\right)q_{TC,2} + \left((q_{E,1} - 2K_{Q0}N_0)q_{TC,0} - q_{TC,1}q_{E,0}\right)k_2} \\ \dfrac{-N_0^2 \cdot \left(q_{TC,1} + k_1 q_{TC,0}\right) \cdot k_Q\left(t \to \infty\right)}{\left(q_{E,1} - 2K_{Q0}N_0 + q_{E,0}k_1\right)q_{TC,2} + \left((q_{E,1} - 2K_{Q0}N_0)q_{TC,0} - q_{TC,1}q_{E,0}\right)k_2} \end{bmatrix} \tag{6.74}$$

It is evident from the above that a major source of steady-state error, in the framework of UPM and NPM, is modification of propeller-law coefficient K_Q, *i.e.* any permanent deviation from nominal value K_{Q0}. This can happen if propulsion power requirements are increased. In turn, ship propulsion power requirements may be increased due to hull and/or propeller fouling, loading or operating (weather/sea) conditions, *etc*. For example, after a long period in service without drydocking hull fouling is observed due to pollution. This is translated to increased power requirements in order to maintain the same service speed as before, as the resistance force to vessel advance is increased. In effect, and taking into account the considerations of Chapter 1, the propeller torque demands are increased, resulting in an increased value of K_Q.

Another source of steady-state error in state variables is reduction of engine and/or turbocharger efficiency due to powerplant performance degradation and/or ageing (*e.g.* piston ring wear or fouling of turbine wheel). Indeed, ideally it holds that:

$$Q_E\left(N_0, N_{TC0}, F_{R0}\right) = Q_{E0} = K_{Q0} \cdot N_0^2 \tag{6.75}$$

However, due to degradation, it can happen that at one or more N^{2}M steady-state operating points:

$$Q_E\left(N_0, N_{TC0}, F_{R0}\right) = Q'_{E0} < Q_{E0} = K_{Q0} \cdot N_0^2 \tag{6.76}$$

and/or

$$Q_{TC}\left(N_0, N_{TC0}, F_{R0}\right) = Q'_{TC0} \neq 0 \tag{6.77}$$

Using linearisation analysis we obtain that:

$$\begin{aligned} Q_E\left(N, N_{TC}, F_R\right) &= Q'_{E0} + q_{E,0} \cdot f_R \\ Q_{TC}\left(N, N_{TC}, F_R\right) &= Q'_{TC0} + q_{TC,2} \cdot n_{TC} + q_{TC,0} \cdot f_R \end{aligned} \tag{6.78}$$

In the above, to simplify the expressions and without any loss of generality, only the dominant terms are kept, by taking into account the facts for the values of torque derivatives valid for common marine plants, presented in Section 5.4.2.

The steady-state error in the state vector $\mathrm{x}_{\mathrm{SS}}^{\mathrm{T}} = \begin{bmatrix} n_{\mathrm{SS}} & n_{TC,\mathrm{SS}} \end{bmatrix}$ can be calculated by using the above linear approximations, in order that shaft equilibrium is restored for the closed-loop plant with linear state-feedback controls of the form $f_R = k_1 \cdot n + k_2 \cdot n_{TC}$:

$$\begin{Bmatrix} Q'_{E0} + q_{E,0}k_1 \cdot n_{SS} + q_{E,0}k_2 \cdot n_{TC,SS} = Q_{E0} + 2K_{Q0}N_0 \cdot n_{SS} \\ Q'_{TC0} + q_{TC,2} \cdot n_{TC,SS} + q_{TC,0}k_1 \cdot n_{SS} + q_{TC,0}k_2 \cdot n_{TC,SS} = 0 \end{Bmatrix}$$

$$\Updownarrow$$

$$\mathbf{x}_{SS} = \begin{bmatrix} n_{SS} \\ n_{TC,SS} \end{bmatrix} = \begin{bmatrix} q_{E,0}k_1 - 2K_{Q0}N_0 & q_{E,0}k_2 \\ q_{TC,0}k_1 & q_{TC,0}k_2 + q_{TC,2} \end{bmatrix}^{-1} \cdot \begin{bmatrix} Q_{E0} - Q'_{E0} \\ -Q'_{TC0} \end{bmatrix} \tag{6.79}$$

In the above, no deviation from the nominal propeller law is assumed (*i.e.* $K_Q = K_{Q0}$) in order to examine the isolated effect of powerplant degradation on steady-state error; however, the above can be easily extended in order to include the effect of $k_Q(t \to \infty)$, as done previously.

As can be seen, in the open-loop case, *i.e.* $k_1 = k_2 = 0$, the off-diagonal elements are zero. Therefore, engine torque deviation from nominal is translated only to shaft rpm deviation and turbocharger torque deviation is translated only to turbo rpm deviation, meaning that the two states are decoupled. On the other hand, state-feedback control allows redistribution of steady-state error, due to deviation of one of the torque variables, to both state variables. The same holds in the case where $k_Q(t \to \infty) \neq 0$, as can be seen from the expressions obtained previously; specifically, when no feedback control is applied, a deviation from the nominal propeller law affects mainly shaft rpm and not turbo rpm, as the value of parameter $q_{TC,1}$ is usually small in two-stroke engines.

Occurrence of steady-state error in one or both state variables, and, in effect, a steady-state nonzero value for incremental control action f_R, manifests itself in permanent mismatch between actual plant operation and ideal behaviour, quantified by N^2M. Detection of such a mismatch may prove useful in automated monitoring systems, in order to detect anomalies and the need for maintenance. Alternatively, if zero steady-state error is to be re-established, recalibration of N^2M is required in order for the model to be updated with the modification of the various system parameters. This means that the new value (or set of values) for nominal propeller torque coefficient needs to be determined and/or the neural torque approximators need to be retrained using current engine time series.

6.4.2 Integral Control and Steady-state Error

As early as in Section 4.1.3 it was mentioned that the incorporation of an integral term in the control law allows for elimination of steady-state error. For example, in the case of the PID control law, this has been adequately demonstrated. However, the approach of PID control for marine plants is based on scalar transfer functions. Even in this case, though, it has been shown that the order of the closed-loop system is increased by one when the shaft rpm integral is propagated through the control law to the control action.

In the more general case where a MIMO state-space description of the open-loop plant is available, instead of a scalar transfer function, the state vector has to be enriched with the appropriate integrals of the "real" state variables that are of interest. This technique is exemplified in the case of the marine plant state equations. Specifically, the state vector is extended in order to include the following:

$$x_3(t) = \int_0^t n(\xi)d\xi \text{ and } x_4(t) = \int_0^t n_{TC}(\xi)d\xi \tag{6.80}$$

The following extended open-loop state equations are then formulated:

$$\begin{bmatrix} \dot{x}_1 \\ \dot{x}_2 \\ \dot{x}_3 \\ \dot{x}_4 \end{bmatrix} = \begin{bmatrix} a_{11} & a_{12} & 0 & 0 \\ a_{21} & a_{22} & 0 & 0 \\ 1 & 0 & 0 & 0 \\ 0 & 1 & 0 & 0 \end{bmatrix} \cdot \begin{bmatrix} x_1 \\ x_2 \\ x_3 \\ x_4 \end{bmatrix} + \begin{bmatrix} b_1 \\ b_2 \\ 0 \\ 0 \end{bmatrix} \cdot f_R + \begin{bmatrix} d_1 \\ 0 \\ 0 \\ 0 \end{bmatrix}$$

$$\Updownarrow$$

$$\dot{\hat{\mathrm{x}}} = \hat{\mathrm{A}} \cdot \hat{\mathrm{x}} + \hat{\mathrm{B}} \cdot f_R + \hat{\mathrm{d}} \tag{6.81}$$

Then, full-state-feedback control is applied to the above extended ("hat") system:

$$f_R = \hat{\mathrm{K}} \cdot \hat{\mathrm{x}} = \begin{bmatrix} \hat{k}_1 & \hat{k}_2 & \hat{k}_3 & \hat{k}_4 \end{bmatrix} \cdot \begin{bmatrix} x_1 & x_2 & x_3 & x_4 \end{bmatrix}^{\mathrm{T}} \tag{6.82}$$

Therefore the following closed-loop transfer function from disturbance to the state is obtained:

$$\hat{\mathrm{x}}(s) = (s\mathrm{I}_4 - \hat{\mathrm{A}} - \hat{\mathrm{B}} \cdot \hat{\mathrm{K}})^{-1} \cdot \hat{\mathrm{d}}(s) \tag{6.83}$$

As disturbance is actually scalar, in effect four scalar transfer functions from d_1 to the four state variables have to be shaped; moreover, the characteristic polynomial of all these functions will have order four. By substituting the "dummy" variables in the control law equation it is obvious that integrals of the two state variables, namely x_1 and x_2, in question will appear. Consequently, it can be expected that the closed-loop system will obtain zeros at $s = j\omega = 0$, meaning that steady-state error is eliminated. Furthermore, additional constraints concerning the magnitude of the integrals can be satisfied by appropriately bounding the H∞-norm of the transfer functions from disturbance to x_3 and x_4 and calculating the corresponding controller gains. However, in order for this to be achieved one has to calculate the roots of a fourth-order polynomial; this makes the design process rather cumbersome and promotes the standard H∞ synthesis procedure probably as the only practically tractable solution.

Alternatively, if the integral of only one state variable, most probably shaft rpm, is introduced to the control law the order of the characteristic polynomial is reduced to three. Although the other state variable (the one whose integral is not in the control law) will still exhibit steady-state error, this compromise is still attractive because there is an analytical method for the determination of the location of the roots of third-order polynomials on the s-plane. The method was originally proposed by Vishnegradskii [72], concerning asymptotic stability of third-order characteristic polynomials, according to the values of their coefficients, and then was extended by Aizerman [73,74] in order to assess the relative stability of a system. It should be pointed out that the parameters involved in the method include a damping factor ζ, and consequently it comes as a natural extension of the analysis for second-order systems presented in detail in Chapter 4.

6.5 Summary

The powerplant model decomposition, obtained in Chapter 5, is employed in order to propose a propulsion control architecture partitioned in order to provide a reference value set for the idealised operation of the plant, in combination with compensation to deviations from this optimised behaviour. In effect, N^2M is proposed to form the core of a supervisory controller module that is capable of generating setpoint values for the feedback compensatory control. The importance of filtering is pointed out for this part of propulsion control, which, in combination with the non-linear nature of this plant model, allows for improvement in transient response and smoother engine running. Then the feedback control part is examined in the framework of modern theory of robust control for linear systems. The drawbacks of the standard techniques are examined and then a controller design method, tailored for marine plants is proposed. The proposed method is actually individual for each state variable closed-loop shaping, so that practical, separate H_∞-norm requirements are met. Considerations for robustness against state-space parametric uncertainty naturally arise, and the applicability of standard theoretical tools is directly examined and proposed. Finally, steady-state error analysis is given in combination with integral control.

CHAPTER 7

CLOSURE

7.1 Conclusions and Discussion

The practices, techniques and results used and obtained in this work are briefly commented and discussed here.

The quasi-steady cycle-mean model for the engine thermodynamic processes, employed in this work, has been used extensively and is widely validated into literature for energetic and power calculations, as well as for turbocharger matching. Furthermore, although the physical equations are presented in a somewhat *ad hoc* manner in the text, the physical background and motivation is rather strong, as can be seen in typical thermodynamics textbooks. The solution procedure is actually typical for non-linear, perplexed algebraic sets of equations and the usage of maps is common in the reciprocating engine industry, not only in the case of torques but for other variables as well. Finally, the assumption of modelling shafts as lumped mechanical elements is encountered quite often in literature.

In Chapter 3, the first approach to plant modelling for control is covered in detail. Input–output, black-box modelling, based on the results and insight provided by the quasi-steady, cycle-mean thermodynamic analysis, provided the scalar transfer function of the marine plant. Although a number of transfer functions for the marine plant are also presented in literature, the use of thermodynamic filling-and-emptying engine models for obtaining values of the coefficients of the open-loop plant transfer function is proposed. As explained, use of such models can provide additional data series to manufacturer steady-state performance curves without the need for shipboard measurement campaigns and open- or closed-loop experiments, which, in many cases are, besides being costly, are rather difficult to implement. Finally, linear analysis of the marine shafting system is done, in order to justify the assumptions made for controller design and provide the basis for engine transient control utilising the shaft torque feedback signal.

The empirical transfer function obtained for the marine plant is employed in Chapter 4 in order to derive design constraint equations for the PID control law, which is encountered most often in marine and industrial practice. After a brief review of the various forms of the PID control law, in relation to their drawbacks and virtues, a synthesis technique based on the notion of $H\infty$-norm and aiming to reduce the propeller load disturbance on normal plant operation is deployed. The method is applied for both PI and PID compensation. However, the introduction of fast, higher-order dynamical terms has revealed that the PI scheme fails to demonstrate the required robustness. On the other hand, the PID scheme demonstrates this form of robustness, but, then, a manner for overcoming the practical noise problems appearing when D-term based control is employed is required. The effect of the process on noise is also investigated, which also leads to the same conclusion, *i.e.* that a noise-immune implementation of the D-term is needed. This is achieved by employing the shaft torque signal and using the shafting system dynamical analysis results. In effect, a control scheme with both increased robustness against neglected dynamical terms, as well as process and measurement noise immunity is proposed based on shaft rpm feedback, complemented by shaft torque feedforward for avoiding real-time rpm signal differentiation, which amplifies high-frequency noise.

In Chapter 5, the first step towards the application of modern state-feedback control is made by the development of the state-space description of the open-loop propulsion plant. The approach is based on the torque maps obtained by the numerical solution of the algebraic set of equations used in Chapter 2. These maps are used for training and validating standard feedforward three-layered neural nets aiming to approximate these two torque maps with required accuracy. Based on standard theoretical results concerning both non-linear mapping ability and backpropagation training convergence for three-layered neural nets, the final outcome is two neural torque approximators: one for engine torque delivery and one for turbocharger total accelerating torque. These approximators prove rather useful for further control developments mainly because they are differentiable and allow for derivative torque functions to be generated over the entire operating range of interest for the plant. In effect, by using the approximator functions, the plant description is decomposed to a nominal non-linear model, representing the desired behaviour, and an uncertain perturbation model, derived by linearisation. Dynamics of N^2M are clearly formulated using the neural approximators. On the other hand, approximator partial derivative functions allow for the estimation of bounds for the uncertainty inherent in UPM, as well as for the propeller load disturbance. This is done by injecting the established bounds for the uncertain physical parameters in the state-space model parameters; thus the nonlinearity and complexity of propulsion powerplant processes and operation is substituted by a nominal non-linear model for ideal behaviour, and a family of linear perturbation models.

Finally, in Chapter 6 an alternative to the standard $H\infty$ state-feedback controller synthesis is proposed for the case of marine plants; however, the method is also applicable to systems with same structure and a single, scalar source of disturbance. The method dispenses with the need to calculate and design the $H\infty$-norm of the closed-loop transfer function matrix by substituting it with separate scalar $H\infty$-norm requirements. Hence, the design provides more insightful results and allows for the reduction of numerical methods in favour of analytical ones, which can be manipulated more insightfully. Finally, a proposal for an open-loop controller based on the neural torque approximators that actually depicts the desired behaviour of the marine plant and can be recalibrated when needed then covers the lack of operating point control.

7.2 Subjects for Future Investigations and Research

As has been seen in the text, there are three clearly distinguished threads for future investigations: (a) experimental validation and shipboard implementation of the proposed designs; (b) propulsion plant modelling for control; (c) engine and propulsion controller synthesis and design.

Thread (a) is straightforward, and should it be effectuated, may come up with marine propulsion powerplants of enhanced safety and reliability, as well as with increased margins and tolerances, combined with reduced overhead installation and running costs.

Concerning (b), the following are some topics that may be included in future research:

- Development of a methodology for real-time and possibly online identification of the powerplant "nominal", model embedded in the supervisory controller module, in order that the effect on performance of parameters, such as ambient conditions, lubrication, *etc.*, as well as plant and/or hull ageing and degradation, is estimated more accurately.
- Inclusion of the dynamic effect of auxiliary systems, such as starting air and auxiliary blowers, especially during rapid engine loading or unloading (*e.g.* crash stop or emergency reverse), and allowing for a "nominal" model and a supervisory controller that can deal with these situations.
- Interconnection of propeller load demand with ship motions including, besides of forward advance, roll, pitch, heave, *etc.*, allowing for real-time assessment of severity of propeller load demand fluctuation, in order for model-based adaptive propulsion control schemes to be examined.

Finally, thread (c) of controller synthesis could focus on the following research fields:

- Application of the formal $H\infty$ synthesis procedure and comparison of results with those obtained in this work, combined with extension of the proposed methods in order to include sensor/actuator dynamics and measurement noise.
- Combination of full-state-feedback control with integral control in order to eliminate steady-state error of one or both state variables, from the perspective presented in Chapter 6.
- Provided that a dynamic model for the interactions between ship motions and propeller load torque has been established, model-based

propulsion control strategies, combining robust and adaptive control theoretical tools can be investigated.

APPENDIX A
NON-LINEAR ALGEBRAIC SYSTEMS OF EQUATIONS

A straightforward numerical solution procedure of algebraic systems of the form:

$$\vec{\theta}(\vec{\xi}) = \vec{0} \tag{A.1}$$

is derived from the Newton–Raphson method in one dimension. The *k*th iterative step in the multidimensional Newton–Raphson method is as follows [13]:

$$\vec{\xi}^{<k>} = \vec{\xi}^{<k-1>} - (\vec{\theta}')^{-1} \cdot \vec{\theta}(\vec{\xi}^{<k-1>}) \tag{A.2}$$

where:

$$\vec{\theta}' = \left[\left. \frac{\partial \theta_i}{\partial \xi_j} \right|_{\vec{\xi}^{<k-1>}} \right] \tag{A.3}$$

However, the Newton–Raphson method in both one or more dimensions suffers from the major problem of poor convergence in early iterative steps combined with the hazard of failing if a function extremum (minimum or maximum) is encountered as the derivative is zeroed. Moreover, in [13,14] it is argued that, if the Newton–Raphson method is applied, then by using a numerical difference to approximate the derivative, additional problems arise. Due to the complexity of functions $F_{\text{AIR}}(\dot{m}_A, p_E)$ and $F_{\text{EXH}}(\dot{m}_A, p_E)$ involved in the algebraic system of interest, numerical difference approximation, instead of analytic derivative calculation, is highly preferable. Therefore, the multidimensional Newton–Raphson method is not the solution procedure of choice.

Alternatively, numerical solution of the algebraic system consisting of groups A.1 and A.it can be dealt with as a minimisation (or equivalently optimisation) problem [13,14]. Indeed, the problem of solution is transformed to a minimisation problem as follows.

Based on the ν equations of the perplexed, non-linear algebraic system:

$$\vec{\theta}(\vec{\xi}) = \vec{0} \Leftrightarrow \begin{cases} \theta_1(\xi_1, \xi_2, \ldots, \xi_\nu) = 0 \\ \theta_2(\xi_1, \xi_2, \ldots, \xi_\nu) = 0 \\ \vdots \\ \theta_\nu(\xi_1, \xi_2, \ldots, \xi_\nu) = 0 \end{cases} \tag{A.4}$$

the following objective (cost) function is established:

$$\Theta(\xi_1, \xi_2, \ldots, \xi_\nu) = \sum_{i=1}^{\nu} \theta_i^2(\xi_1, \xi_2, \ldots, \xi_\nu) \tag{A.5}$$

Optimisation problems with the above form of objective function are quite often referred to in related optimisation literature as "non-linear least squares" problems. Note that a solution to the above non-linear least squares problem is not necessarily a solution of the associated algebraic system. Indeed, assuming that the ν-ad $(\xi_1^*,\xi_2^*,\ldots,\xi_\nu^*)$ minimises the objective function $\Theta(\xi_1,\xi_2,\ldots,\xi_\nu)$ to a value $\Theta^* \neq 0$, then, evidently, $(\xi_1^*,\xi_2^*,\ldots,\xi_\nu^*)$ is not satisfying the algebraic system $\vec{\theta}(\vec{\xi})=\vec{0}$. However, candidate solutions of $\vec{\theta}(\vec{\xi})=\vec{0}$ can be sought in the set of (local) minima of objective function $\Theta(\xi_1,\xi_2,\ldots,\xi_\nu)$.

As the functions $\theta_i\ (\xi_1,\xi_2,\ldots,\xi_\nu)$, $i=1,2,\ldots,\nu$ are non-linear, steepest-descent algorithms were not considered for the solution of the associated minimisation problem. On the other hand, performance of medium- and large-scale minimisation (optimisation) algorithms has been investigated. The Levenberg–Marquardt method is one of the medium-scale techniques that can be employed [13,15]. The essentials of this technique were presented in Section 5.2.2 when the methods for determining the weights of a neural net are surveyed. Here, it is said that it has been proven to perform rather poorly in the case of the thermodynamic engine model algebraic system. Therefore, large-scale algorithms were used, and specifically the trust region method [13].

The trust region approach to minimisation (optimisation) is based on the idea of approximating the function with a simpler one, which reasonably reflects the behaviour of the original function in the neighbourhood around a specified point. This neighbourhood is the trust region. A trial step is computed by minimising (or to be more accurate approximately minimising) over the trust region. This is the trust region subproblem.

The key questions in defining a specific trust region strategy to minimising a function are how to choose and compute the approximating simpler function (defined at the current point), how to choose and modify the trust region, and how accurately to solve the trust region subproblem. In the standard trust region method, the quadratic approximation function is used by keeping the first two terms of the function's Taylor series expansion; the neighbourhood is usually spherical or ellipsoidal in shape. Mathematically the solution to the trust region subproblem requires second-order information as, besides the gradient of the objective function at the current "central" point, the Hessian matrix (the symmetric matrix of second derivatives) must also be evaluated.

APPENDIX B

SECOND-ORDER TRANSFER FUNCTIONS WITH ZERO

B.1 Transient Behaviour Analysis

The following second-order transfer function is of major concern for the development of PID control for marine plants:

$$G_c(s) = \frac{K\omega_n^2 \cdot s}{s^2 + 2\zeta\omega_n \cdot s + \omega_n^2} \tag{B.1}$$

Extensive results and analyses can be found in literature, *e.g.* in [10,33], for the following standard form of the second-order transfer function:

$$G_{c,\dagger}(s) = \frac{\omega_n^2}{s^2 + 2\zeta\omega_n \cdot s + \omega_n^2} \tag{B.2}$$

Some of the most important conclusions of the analysis of $G_c(s)$ are given briefly here. Most of them agree with those for $G_{c,\dagger}(s)$. Note that parameter ω_n (ω_n is assumed always positive with no loss of generality) is referred to as natural underdamped frequency, ζ as damping ratio and K as steady-state gain.

The most important parameter of the above, which determines the roots of the characteristic polynomial (denominator polynomial) $p_c(s) = s^2 + 2\zeta\omega_n \cdot s + \omega_n^2$ of both the above transfer functions, is damping ratio ζ. As explained in [10,33], the following cases can be distinguished according to the value of ζ:

(1) $\zeta < 0$. The above transfer function has two poles in the right half of the *s*-plane (positive real part). Therefore, the system is unstable. This case will not be considered any further as it is of no practical interest.

(2) $\zeta = 0$. The transfer function has two conjugate, purely imaginary poles (located on the $j\omega$-axis). Therefore, the system is stable but not asymptotically stable, *i.e.* it generates undamped sinusoidal oscillations as a response to a step input. The system poles are given by the following:

$$s_{1,2} = \pm j\omega_n \tag{B.3}$$

However, this case is also of no practical interest for marine plants and will not be considered any further.

(3) $0 < \zeta < 1$. The transfer function has two, distinct complex poles located in the left half of the s-plane (negative real parts). The two system poles are:

$$s_{1,2} = -\omega_n \zeta \pm j\omega_n \cdot \sqrt{1-\zeta^2} \qquad \text{(B.4)}$$

The step response of a system with such a characteristic polynomial may exhibit oscillation, and therefore the system is underdamped. As a consequence, step response is rather fast but exhibits overshoot, which is a function of ζ.

(4) $\zeta = 1$. The transfer function has a single, double real pole located on the negative real axis. Therefore, the system is asymptotically stable; additionally, as the imaginary part of the double pole is zero, the step response exhibits no oscillation and the system is said to be critically damped. The system's double real pole is:

$$s_{1,2} = -\omega_n \qquad \text{(B.5)}$$

(5) $\zeta > 1$. The transfer function has two distinct real poles located in the left half of the s-plane. Therefore, the system is asymptotically stable; additionally, as the imaginary part of the poles is zero, the step response exhibits no oscillation. The two real poles of the system are:

$$s_{1,2} = -\omega_n \zeta \pm \omega_n \cdot \sqrt{\zeta^2 - 1} \qquad \text{(B.6)}$$

Two rather important remarks can be made concerning this case. First, it should be noted that the speed of response is actually determined by the pole closest to $s = 0$, i.e.:

$$s_1 = \omega_n \cdot \left(-\zeta + \sqrt{\zeta^2 - 1}\right) \qquad \text{(B.7)}$$

As can be seen from Figure B.1, which shows a plot of function $s_1 / \omega_n = -\zeta + \sqrt{\zeta^2 - 1}$ for values of ζ in the interval [1,3], the "slow" system pole s_1 is increasing (decreasing in absolute value) for increasing values of ζ.

Therefore, system response is growing slower in comparison to the critically damped case (4), assuming the same value for ω_n. Additionally, for values of $\zeta \gg 1$:

$$s_1 \to 0^- \qquad \text{(B.8)}$$

Therefore, the system, besides becoming slower, is also exhibiting reduced relative stability, compared to the critically damped case.

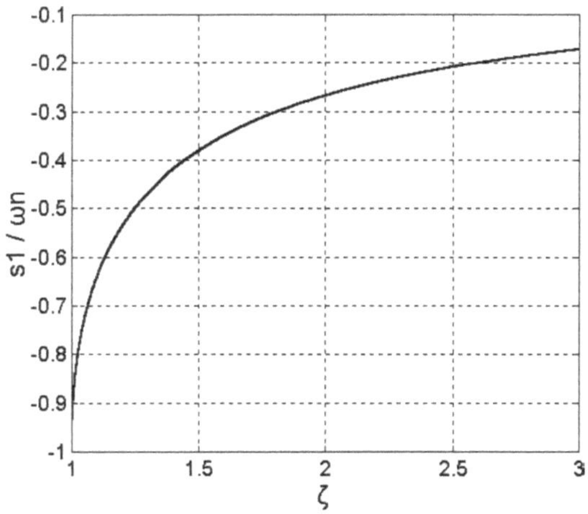

Figure B.1 Function $s_1 / \omega_n = -\zeta + \sqrt{\zeta^2 - 1}$ for ζ in the interval $[1,3]$

As already mentioned, the main differences between transfer functions $G_c(s)$ and $G_{c,\dagger}(s)$ are the introduction of a zero at $s = 0$ and a steady-state gain K.

The zero at $s = 0$ causes the steady-state response of $G_{c,\dagger}(s)$ to become:

$$G_c(s = 0 + j0) = 0 \text{ whilst } G_{c,\dagger}(s = 0 + j0) = 1 \tag{B.9}$$

The overall effect of introducing the zero at $s = 0$, and the non-unity steady-state gain K can be understood by noticing that:

$$G_c(s) = Ks \cdot G_{c,\dagger}(s) \Rightarrow g_c(t) = K \cdot \frac{d}{dt} g_{c,\dagger}(t) \tag{B.10}$$

where $g_c(t)$ and $g_{c,\dagger}(t)$ are the unit step responses corresponding to transfer functions $G_c(s)$ and $G_{c,\dagger}(s)$ respectively.

The application of the above general relation to the case of sinusoidal steady-state response yields:

$$g_{c,\dagger}(t) = \sin(\omega t) \Rightarrow g_c(t) = K\omega \cdot \cos(\omega t) \tag{B.11}$$

Therefore, differentiation introduces a phase shift of $(\pi / 2)$, as well as a linear dependence of the amplitude on frequency ω. Sinusoidal steady-state response is examined closer below, using the Bode plots, in conjunction with the notion of H∞-norm of a scalar transfer function.

B.2 Frequency Response and H∞-norm Requirements

Transient performance in disturbance rejection can be quantified using the H∞-norm of a scalar transfer function. The H∞-norm is actually the peak of the magnitude Bode plot of the transfer function. Specifically, given a transfer function $G(s)$ and the corresponding frequency response $G(s = j\omega) = G(j\omega)$, the H∞-norm, $\|G(s)\|_{\infty}$, is defined as follows:

$$\|G(s)\|_{\infty} = \sup_{\omega \geq 0} |G(j\omega)| \tag{B.12}$$

Frequency response is quite commonly used for stating transient performance requirements for linear time-invariant systems. For example, if transfer function $G(s)$ represents the closed-loop transfer function from a reference signal $u_{REF}(s)$ to a controlled output signal $y(s)$ the requirements can be of the form:

$$|G(j\omega)| = 1 \text{ and } \angle G(j\omega) = \tau_{shift} \cdot \omega \tag{B.13}$$

for a specific frequency interval: $0 \leq \omega_1 \leq \omega \leq \omega_2$. The above is the standard specification of an all-pass filter with unity gain for the band $[\omega_1, \omega_2]$. Alternatively, H_2-norm requirements and specifications can be employed for the formulation of the so-called tracking problem.

Another possibility for $G(s)$, however, is that it represents the closed-loop transfer function from a disturbance signal $d(s)$ to a controlled output $y(s)$. In that respect, requirements for $G(s)$ in the frequency domain concerning the H∞-norm, $\|G(s)\|_{\infty}$, can help in the formulation of the feedback control problem. Specifically, some knowledge of the spectral content of $d(s)$ is usually available. Then, for the frequency band(s) for which the psd or, equivalently, the magnitude $|d(j\omega)|$ of the disturbance $d(s)$ is not negligible, a requirement of the following form suffices:

$$\sup_{\omega_1 \leq \omega \leq \omega_2} |G(j\omega)| \leq G_0 \tag{B.14}$$

However, in most cases it is desirable to extend the above to the entire frequency range $\omega \geq 0$. Indeed, significant insight and mathematical compactness is gained in that way. Additionally, the extended requirement can cope with the more general case according to which no detailed knowledge of the spectral content of the disturbance signal involved is available; actually, only the magnitude peak value over the entire frequency range is required. From this perspective, the above "practical" requirement can be expanded as follows:

$$\|G(s)\|_{\infty} = \sup_{\omega \geq 0} |G(j\omega)| \leq G_0 \tag{B.15}$$

In the case of marine plants, disturbance rejection is a major control problem, especially during operation near MCR and under certain weather/sea conditions introducing increased propeller load torque fluctuations from the nominal propeller curve values. Furthermore, as demonstrated in the Section 4.3.2, the closed-loop transfer function from the disturbance to the shaft rpm perturbation signal takes the form of the second-order transfer function with zero, presented in Section B.1

(Equation (B.1)). Therefore, the frequency response of the transfer function $G_c(s)$:

$$G_c(s) = \frac{K\omega_n^2 \cdot s}{s^2 + 2\zeta\omega_n \cdot s + \omega_n^2} \tag{B.16}$$

is investigated in the rest of this section. The analysis is focused on the H∞-norm $\|G_c(s)\|_\infty$, as shaft rpm (speed) regulation is a core subject of this work.

As already mentioned, the behaviour of $G_c(s)$ is mainly determined by the value of damping ratio ζ. Therefore, typical (magnitude and phase) Bode plots for cases (3), (4) and (5) of the Section B.1 are given in Figure B.2. The first 10 s of the step response of the systems is also plotted. The natural underdamped frequency ω_n is 1.0 rad/s for all cases; the steady-state gain K is 1.0 in all cases, too. The three transfer functions considered as examples are summarised in Table B.1.

Table B.1 Typical second-order transfer functions with zero at $s = 0$

ζ	Transfer function	Poles
0.4	$G_c(s) = \dfrac{s}{s^2 + 0.8s + 1}$	$s_{1,2} = -0.4 \pm 0.9165i$
1.0	$G_c(s) = \dfrac{s}{s^2 + 2s + 1}$	$s_{1,2} = -1.0$
3.0	$G_c(s) = \dfrac{s}{s^2 + 6s + 1}$	$s_1 = -0.1716$, $s_2 = -5.8284$

Note that cases for which $G_c(s)$ is not asymptotically stable are excluded from the frequency response investigations. On the other hand, transfer function:

$$G_{c,\dagger}(s) = \frac{1}{s^2 + 2s + 1}$$

is also included in Figure B.2. This is the counterpart of the case with $\zeta = 1.0$ but without a zero at $s = 0$. The objective is to demonstrate the effect of introducing the zero and to examine how the step and steady-state sinusoidal response are modified.

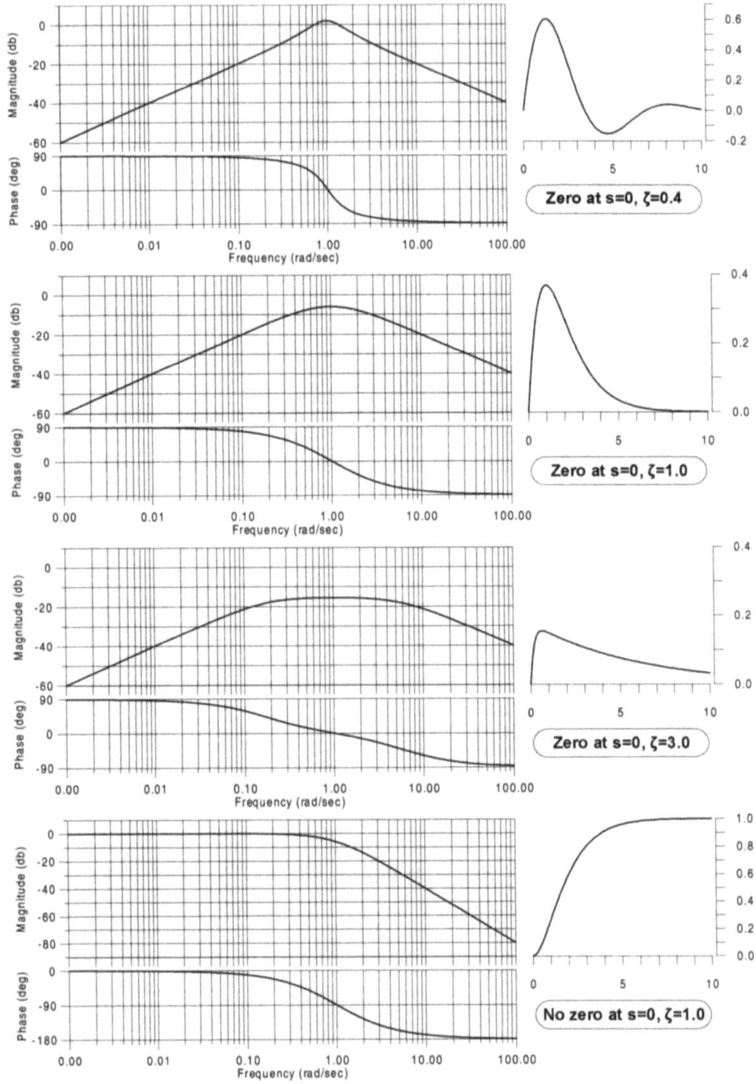

Figure B.2 Bode plots and step responses of typical second-order transfer functions

Some remarks can be made concerning the plots in Figure B.2. First, in both cases (3) and (4) (*i.e.* for $\zeta \leq 1.0$) the peak of the magnitude Bode plot is located at $\omega = \omega_n$. Therefore:

$$\zeta \leq 1.0 \Rightarrow \|G_c(s)\|_\infty = |G_c(j\omega_n)| \tag{B.17}$$

Additionally, it is observed that the H∞-norm, $\|G_c(s)\|_\infty$, is an increasing function of the damping ratio ζ.

For case (4), $\zeta = 1.0$, and it can be calculated that:

$$\left\|G_c(s)\right\|_\infty = \left|G_c(j\omega_n)\right| = \left|\frac{K\omega_n}{2}\right| = \frac{|K|\omega_n}{2} \tag{B.18}$$

This result is used extensively during the synthesis of the PID controller.

Further increase of ζ, beyond unity, is translated to reduction of $\left\|G_c(s)\right\|_\infty$. However, as demonstrated in Figure B.2, this reduction is combined with significant "flattening" of the magnitude plot. Indeed, the magnitude plot demonstrates a "plateau" in the frequency range defined by the inequalities:

$$|s_1| = \omega_n \cdot \left(\zeta - \sqrt{\zeta^2 - 1}\right) \le \omega \le \omega_n \cdot \left(\zeta + \sqrt{\zeta^2 - 1}\right) = |s_2| \tag{B.19}$$

ω_n is also located inside the above frequency range for damping ratio values $\zeta > 1.0$. Indeed, it holds that:

$$|s_1| = \omega_n \cdot \left(\zeta - \sqrt{\zeta^2 - 1}\right) \le \omega_n \le \omega_n \cdot \left(\zeta + \sqrt{\zeta^2 - 1}\right) = |s_2|$$

The rightmost inequality is obvious, whilst the leftmost can be deduced by using the plot of the function $s_1 / \omega_n = -\zeta + \sqrt{\zeta^2 - 1}$ given in Figure B.1. As can be seen:

$$-1 \le \frac{s_1}{\omega_n} < 0 \Rightarrow \left|\frac{s_1}{\omega_n}\right| \le 1, \text{ for } \zeta \ge 1 \tag{B.20}$$

However, the positive effect on $\left\|G_c(s)\right\|_\infty$ introduced by increasing ζ beyond unity has to be balanced by considering the relative stability compromise. Indeed, for $\zeta \ge 1$, dominant pole $s_1 = \mathrm{Re}\{s_1\} \to 0^-$ for $\zeta \to +\infty$. Moreover, note that:

$$s_2 = \mathrm{Re}\{s_2\} \to -\infty \text{ for } 1 \le \zeta \to +\infty \tag{B.21}$$

Both the above results are used extensively for the design of the PID marine engine speed governor.

Before concluding this section, the effect of the zero at $s = 0$ present in transfer function $G_c(s)$, whilst missing in $G_{c,\dagger}(s)$, is investigated. As already mentioned, the major result is for the value $\left|G_c(j0)\right|$ which is zero, whilst $\left|G_{c,\dagger}(j0)\right| = 1 \ne 0$. Moreover, as can be seen from the Bode plots of $G_c(s)$ and $G_{c,\dagger}(s)$ for values of damping ratio $\zeta = 1.0$ and natural underdamped frequency $\omega_n = 1.0$ rad/s, the H∞-norm is reduced by the introduction of the zero at $s = 0$. A more important consequence is that psd as a function of ω is significantly smaller in the low-frequency range, where usually the spectral content of the involved disturbance $d(s)$ is concentrated (at least in the case of marine plants). Indeed, it can be seen from Figure B.2 in the case for $\zeta = 1.0$, that:

$$\left|G_c(j\omega)\right| < \left|G_{c,\dagger}(j\omega)\right| \Rightarrow \left|G_c(j\omega)\right|^2 < \left|G_{c,\dagger}(j\omega)\right|^2 \text{ for } 0 \le \omega \le \omega_n \tag{B.22}$$

However, the above positive result does not come without a penalty; as demonstrated in the Bode plots, the introduction of zero at $s=0$ causes significant amplitude distortion in the frequency range $[0,\omega_n]$. On the other hand, the magnitude of the frequency response of $G_{c,\dagger}(s)$ is flat in the same frequency range, as it is approximately equal to $\|G_{c,\dagger}(s)\|_\infty$ over the entire band.

The low-frequency amplitude distortion, due to the zero at $s=0$, has an observable effect on the step response of $G_c(s)$. As can be clearly seen, in the Figure B.3 for the step response plots for values of $\zeta=0.4$, 1.0, 3.0 of transfer functions $G_c(s)$ and $G_{c,\dagger}(s)$, rippling is significantly enhanced. This is explained by taking into account that the spectral content of the step signal is concentrated in the low-frequency range. Therefore, significant amplitude distortion in the low-frequency band is manifested by the selectivity imposed on the frequency components.

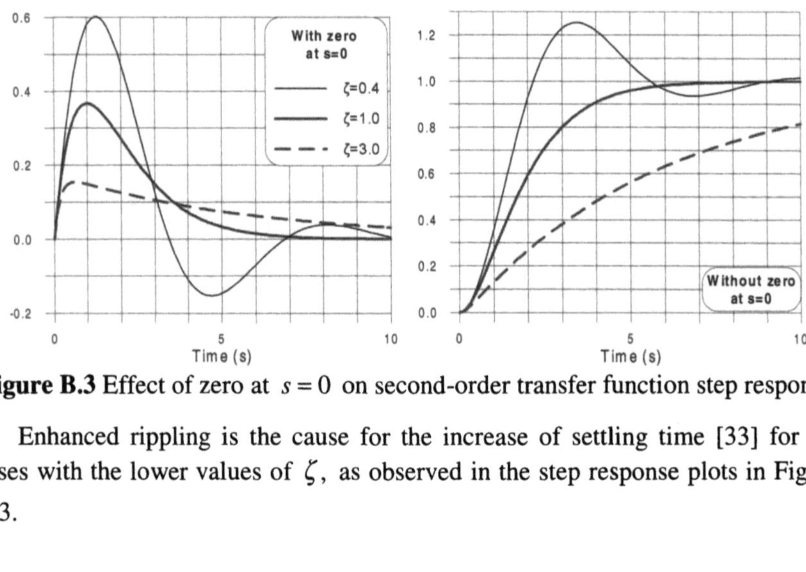

Figure B.3 Effect of zero at $s=0$ on second-order transfer function step response

Enhanced rippling is the cause for the increase of settling time [33] for the cases with the lower values of ζ, as observed in the step response plots in Figure B.3.

REFERENCES

1. Blank DA, Bock AE, Richardson DJ. Introduction to Naval Engineering. US Naval Institute, Annapolis, Maryland, USA, 1985.

2. Heywood JB. Internal Combustion Engine Fundamentals. McGraw-Hill, New York, USA, 1988.

3. Illies K. Low-speed Directly-coupled Diesel Engines. In: Harrington RL (ed.). Marine Engineering. SNAME, USA, 1971; 280-316.

4. Kyrtatos NP, Theotokatos G, Xiros N. Main Engine Control for Heavy Weather Conditions – The ACME Project. In: Proceedings of the 6th International Symposium on Marine Engineering. MESJ, Tokyo, 2000.

5. Sorensen P, Pedersen PS. The Intelligent Engine and Electronic Products – A Development Status. In: 22nd CIMAC Congress. CIMAC, Copenhagen, 1998.

6. Kiencke U, Nielsen L. Automotive Control Systems. Springer, Berlin, Germany, 1999.

7. Meier A. A Simple Method of Calculation for Matching Turbochargers. BBC Brown, Boveri & Company Ltd., Baden, Switzerland; CH-T 120 163 E.

8. Woodward JB, Latorre RG. Modelling of Diesel Engine Transient Behavior in Marine Propulsion Analysis. SNAME Transactions, 1984; 92: 33-49.

9. Leigh JR. Applied Digital Control. Prentice Hall, Great Britain, 1992.

10. Nagrath IJ, Gopal M. Control Systems Engineering. Wiley Eastern, Singapore, 1986.

11. L60MC Mk5 Project Guide. MAN B&W Diesel A/S, Copenhagen, 1996; 1.04.

12. L60MC Mk5 Project Guide. MAN B&W Diesel A/S, Copenhagen, 1996; 2.01-2.24.

13. Press WH, Flannery BP, Teukolsky SA, Vetterling WT. Numerical Recipes. Cambridge University Press, USA, 1986.

14. Coleman T, Branch MA, Grace A. Optimization Toolbox for use with Matlab – User's Guide. The Matworks Inc., 1999.

15. Reklaitis GV, Ravindran A, Ragsdell KM. Engineering Optimization. John Wiley & Sons, USA, 1983.

16. Lan WC, Takesi K, Takesi H. Quasi Steady Simulation of Diesel Engine Transient Performance and Design of Mechatronic Governor. Bulletin of the MESJ, 1996; 24/1: 1-13.

17. Roskilly AP, Mesbahi E. Artificial Neural Networks for Marine System Identification and Modelling. Trans. IMarE, 1996; 108/3: 203-13.

18. Faber EG. Some Thoughts on Diesel Marine Engineering. SNAME Transactions, 1993; 101: 537-82.

19. Faber EG. System Interaction between Hull/Propulsor and Piston Engine/Turbocharger. In: WEGEMT 21st Graduate School. Gerhard-Mercator University, Duisburg, Germany, 1994.

20. Pain HJ. The Physics of Vibrations and Waves. John Wiley & Sons, United Kingdom, 1988.

21. Palm WJ. Modelling, Analysis and Control of Dynamic Systems. John Wiley & Sons, USA, 1983.

22. Rajput BS, Prakash Y. Mathematical Physics. Pragati Prakashan, Meerut, India, 1991.

23. Cheng-Ching Y. Autotuning of PID Controllers. Springer-Verlag, London, Great Britain, 1998.

24. Kiong TK, Qing-Guo W, Chieh HC, Hagglund TJ. Advances in PID Control. Springer-Verlag, London, Great Britain, 1999.

25. Hagglund TJ, Astrom KJ. Automatic Tuning of PID Controllers. In: Levine WS (ed.). Control Engineering Handbook. IEEE Press, USA, 1996; 817-26.

26. Winterbone DE, Thiruarooran C, Wellstead PE. A Wholly Dynamic Model of a Turbocharged Diesel Engine for Transfer Function Evaluation. In: 1977 International Automotive Engineering Congress. SAE, Detroit, 1977.

27. Kyrtatos NP, Theotokatos G, Xiros N. Transient Operation of Large-bore, Two-stroke Marine Diesel Engines – Measurements and Simulations. In: 23rd CIMAC Congress. CIMAC, Hamburg, 2001.

28. Yates GH. Governor Performance Slide Rule. Woodward Governor Nederland B.V., 1977; PMCC 77-11.

29. Carlton JS. Marine Propellers and Propulsion. Butterworth & Heinmann, Oxford, UK, 1994.

30. Roy S, Malik OP, Hope GS. An Adaptive Control Scheme for Speed Control of Diesel driven Powerplants, IEEE Transactions on Energy Conversion, 1991; 6/4: 605-11.

31. K90MC Mk6 Project Guide. MAN B&W Diesel A/S, Copenhagen, 1997; 1.04.

32. K90MC Mk6 Project Guide. MAN B&W Diesel A/S, Copenhagen, 1997; 2.01-2.24.

33. Kuo BC. Automatic Control Systems. Prentice-Hall, New Jersey, USA, 1987.

34. Taylor DA. Marine Control Practice. Butterworths, Essex, England, 1987.

35. Long CL. Propellers, Shafting and Shafting System Vibration Analysis. In: Harrington RL (ed.). Marine Engineering. SNAME, USA, 1971; 362-400.

36. Bose NK. Digital Filters Theory and Applications. North-Holland Elsevier, New York, USA, 1985.

37. Vaidyanathan PP. Design and Implementation of Digital FIR Filters. In: Elliott DF (ed.). Handbook of Digital Signal Processing Engineering Applications. Academic Press, USA, 1987; 55-172.

38. Pashtoon NA. IIR Digital Filters. In: Elliott DF (ed.). Handbook of Digital Signal Processing Engineering Applications. Academic Press, USA, 1987; 289-355.

39. Kyrtatos NP, Theodosopoulos P, Theotokatos G, Xiros N. Simulation of the Overall Ship Propulsion Plant for Performance Prediction and Control. In: MarPower '99 Conference. IMarE, Newcastle-upon-Tyne, UK, 1999.

40. Kyrtatos NP, Politis G, Lambropoulos V, Theotokatos G, Xiros N. Optimum Performance of Large Marine Engines Under Extreme Loading Conditions. In: 22nd CIMAC Congress. CIMAC, Copenhagen, 1998.

41. Ge SS, Lee TH, Wang J. Adaptive Control of Non-affine Nonlinear Systems Using Neural Networks. In: 15th IEEE International Symposium on Intelligent Control. IEEE, Rio, Patras, Greece, 2000.

42. Farrell JA. Neural Control. In: Levine WS (ed.). Control Engineering Handbook. IEEE Press, USA, 1996; 1017-61.

43. Isidori A. Nonlinear Control Systems. Springer-Verlag, London, UK, 1995.

44. Zhou K, Doyle JC. Essentials of Robust Control. Prentice Hall, USA, 1998.

45. Balas GJ, Packard A. The Structured Singular Value (μ) Framework. In: Levine WS (ed.). Control Engineering Handbook. IEEE Press, USA, 1996; 671-87.

46. Chao A, Athans M. Stability Robustness to Unstructured Uncertainty for Linear Time Invariant Systems. In: Levine WS (ed.). Control Engineering Handbook. IEEE Press, USA, 1996; 519-35.

47. Droste W. Dynamic Simulation of Propulsion System. In: WEGEMT 12th Graduate School: Propulsion Systems. London, UK, 1989.

48. Fausett L. Fundamentals of Neural Networks. Prentice-Hall Intl., New Jersey, 1994.

49. Tsoukalas LH, Uhrig RE. Fuzzy and Neural Approaches in Engineering. John Wiley & Sons, USA, 1997.

50. Raisch J, Francis BA. Modeling Deterministic Uncertainty. In: Levine WS (ed.). Control Engineering Handbook. IEEE Press, USA, 1996; 551-60.

51. Houpis CH. Quantitative Feedback Theory Technique. In: Levine WS (ed.). Control Engineering Handbook. IEEE Press, USA, 1996; 701-17.

52. Russell S, Norvig P. Artificial Intelligence A Modern Approach. Prentice Hall, USA, 1995.

53. Rao V, Rao H. C++ Neural Networks and Fuzzy Logic. MIS:Press, New York, USA, 1995.

54. Wang XZ. Data Mining and Knowledge Discovery for Process Monitoring and Control. Springer, UK, 1999.

55. Horn RA. Matrix Analysis. Cambridge University Press, USA, 1985.

56. Barnett S. Matrices Methods and Applications. Oxford University Press, New York, USA, 1990.

57. Lublin L, Athans M. Linear Quadratic Regulator Control. In: Levine WS (ed.). Control Engineering Handbook. IEEE Press, USA, 1996; 635-50.

58. Patek SD, Athans M. MIMO Frequency Response Analysis and the Singular Value Decomposition. In: Levine WS (ed.). Control Engineering Handbook. IEEE Press, USA, 1996; 507-18.

59. Van Spronsen PJ, Tousain RL. An Optimal Control Approach to Preventing Marine Diesel Engine Overloading Aboard Karel Doorman Class Frigates. In: Control Applications in Marine Systems 2001. IFAC, Glasgow, Scotland, 2001.

60. Lublin L, Grocott S, Athans M. H2 (LQG) and H∞ Control. In: Levine WS (ed.). Control Engineering Handbook. IEEE Press, USA, 1996; 651-61.

61. Rantzer A. Stability Conditions for Polytopes of Polynomials. IEEE Transactions on Automatic Control, 1992; 37/1: 79-89.

62. Bose NK, Shi YQ. A Simple General Proof of Kharitonov's Generalised Stability Criterion. IEEE Transactions on Automatic Control, 1987; 34/8: 1233-7.

63. Yeung KS, Wang SS. A Simple Proof of Kharitonov's Theorem. IEEE Transactions on Automatic Control, 1987; 32/9: 822-3.

64. Barmish BR, Shi Z. Robust Stability of a Class of Polynomials with Coefficients Depending Multilinearly on Perturbations. IEEE Transactions on Automatic Control, 1990; 35/9: 1040-3.

65. Cavallo A, Celentano G, De Maria G. Robust Stability Analysis of Polynomials with Linearly Dependent Coefficient Perturbations. IEEE Transactions on Automatic Control, 1991; 36/3: 380-4.

66. Kokame H, Mori T. A Root Distribution Criterion for Interval Polynomials. IEEE Transactions on Automatic Control, 1991; 36/3: 362-4.

67. Soh YC, Foo YK. Generalisation of Strong Kharitonov Theorems to the Left Sector. IEEE Transactions on Automatic Control, 1990; 35/12: 1378-85.

68. Tesi A, Vicino A. Robust Stability of State-space Models with Structured Uncertainties. IEEE Transactions on Automatic Control, 1990; 35/2: 191-8.

69. Saydy L, Tits AL, Abed EH. Guardian Maps and the Generalised Stability of Parameterised Families of Matrices and Polynomials. Math. Control Signals Systems, 1990; 3: 345-71.

70. Bartlett AC, Hollot CV, Lin H. Root Locations of an Entire Polytope of Polynomials: It Suffices to Check the Edges. Math. Control Signals Systems, 1988; 1: 61-71.

71. Tempo R, Blanchini F. Robustness Analysis with Real Parametric Uncertainty. In: Levine WS (ed.). Control Engineering Handbook. IEEE Press, USA, 1996; 495-505.

72. Vishnegradskii I. On Direct Action Regulators. Izv. tekhnol. Inst. St. Petersburg, 1877; 1: 21-64.

73. Aizerman MA. On the Damping of Oscillatory Motion Characterised by a Linear Third Order Equation. Automat. Telemech., Moscow, 1940; 1: 55-65.

74. Aizerman MA. Theory of Automatic Control. Pergamon Press, London, UK, 1963.

INDEX

MIX
Papier aus verantwortungsvollen Quellen
Paper from responsible sources
FSC® C105338

If you have any concerns about our products,
you can contact us on
ProductSafety@springernature.com

In case Publisher is established outside the EU,
the EU authorized representative is:
Springer Nature Customer Service Center GmbH
Europaplatz 3, 69115 Heidelberg, Germany

Printed by Libri Plureos GmbH
in Hamburg, Germany